T0276014

An Introduction to Category Theory

Category theory provides a general conceptual framework that has proved fruitful in subjects as diverse as geometry, topology, theoretical computer science and foundational mathematics. Here is a friendly, easy-to-read textbook that explains the fundamentals at a level suitable for newcomers to the subject.

Beginning postgraduate mathematicians will find this book an excellent introduction to all of the basics of category theory. It gives the basic definitions; goes through the various associated gadgetry, such as functors, natural transformations, limits and colimits; and then explains adjunctions. The material is slowly developed using many examples and illustrations to illuminate the concepts explained. Over 200 exercises, with solutions available online, help the reader to access the subject and make the book ideal for self-study. It can also be used as a recommended text for a taught introductory course.

An Introduction to Category Theory

HAROLD SIMMONS
University of Manchester

CAMBRIDGE
UNIVERSITY PRESS

CAMBRIDGE
UNIVERSITY PRESS

University Printing House, Cambridge CB2 8BS, United Kingdom

Cambridge University Press is part of the University of Cambridge.

It furthers the University's mission by disseminating knowledge in the pursuit of education, learning and research at the highest international levels of excellence.

www.cambridge.org
Information on this title: www.cambridge.org/9780521283045

First published 2011
4th printing 2013

A catalogue record for this publication is available from the British Library

Library of Congress Cataloguing in Publication data
Simmons, Harold
An introduction to category theory / Harold Simmons.
p. cm.
ISBN 978-1-107-01087-1 (hardback)
1. Categories (Mathematics) I. Title.
QA169.S56 2011
512'.62–dc23
2011021721

ISBN 978-1-107-01087-1 Hardback
ISBN 978-0-521-28304-5 Paperback

Additional resources for this publication at www.cambridge.org/simmons

Contents

Preface   *page* vii

1   Categories   1
   1.1   Categories defined   1
   1.2   Categories of structured sets   8
   1.3   An arrow need not be a function   16
   1.4   More complicated categories   27
   1.5   Two simple categories and a bonus   31

2   Basic gadgetry   34
   2.1   Diagram chasing   34
   2.2   Monics and epics   37
   2.3   Simple limits and colimits   43
   2.4   Initial and final objects   45
   2.5   Products and coproducts   47
   2.6   Equalizers and coequalizers   56
   2.7   Pullbacks and pushouts   64
   2.8   Using the opposite category   71

3   Functors and natural transformations   72
   3.1   Functors defined   73
   3.2   Some simple functors   76
   3.3   Some less simple functors   79
   3.4   Natural transformations defined   90
   3.5   Examples of natural transformations   93

4   Limits and colimits in general   108
   4.1   Template and diagram – a first pass   109
   4.2   Functor categories   114
   4.3   Problem and solution   118

4.4	Universal solution	120
4.5	A geometric limit and colimit	124
4.6	How to calculate certain limits	130
4.7	Confluent colimits in *Set*	143
5	**Adjunctions**	**148**
5.1	Adjunctions defined	148
5.2	Adjunctions illustrated	153
5.3	Adjunctions uncoupled	164
5.4	The unit and the counit	170
5.5	Free and cofree constructions	174
5.6	Contravariant adjunctions	186
6	**Posets and monoid sets**	**190**
6.1	Posets and complete posets	190
6.2	Two categories of complete posets	191
6.3	Sections of a poset	193
6.4	The two completions	195
6.5	Three endo-functors on *Pos*	197
6.6	Long strings of adjunctions	199
6.7	Two adjunctions for R-sets	202
6.8	The upper left adjoint	205
6.9	The upper adjunction	209
6.10	The lower right adjoint	212
6.11	The lower adjunction	218
6.12	Some final projects	221
	Bibliography	223
	Index	224

Preface

As it says on the front cover this book is an introduction to Category Theory. It gives the basic definitions; goes through the various associated gadgetry such as functors, natural transformations, limits and colimits; and then explains adjunctions. This material could be developed in 50 pages or so, but here it takes some 220 pages. That is because there are many examples illustrating the various notions, some rather straightforward, and others with more content. More importantly, there are also over 200 exercises. And perhaps even more importantly, solutions to these exercises are available online.

The book is aimed primarily at the beginning graduate student, but that does not mean that other students or professional mathematicians will not find it useful. I have designed the book so that it can be used by a single student or small group of students to learn the subject on their own. The book will make a suitable text for a reading group. The book does not assume the reader has a broad knowledge of mathematics. Most of the illustrations use rather simple ideas, but every now and then a more advanced topic is mentioned. The book can also be used as a recommended text for a taught introductory course.

Every mathematician should at least know of the existence of category theory, and many will need to use categorical notions every now and then. For those groups this is the book you should have. Other mathematicians will use category theory every day. That group has to learn the subject sometime, and this is the book to start that process. Of course, the more advanced topics are not dealt with here.

The book has been developed over quite a few years. Several short courses of about 10 hours have been taught (not always by me) using some of the material. In 2007, 2008, and 2009 I gave a course over the web to about a dozen universities. This was part of MAGIC, the

Mathematics   Access   Grid   Instruction   and   Collaboration cooperative of quite a few University Departments of Mathematics in England and Wales. That was an interesting experience and helped me to split the material into small chunks each of the right length to fit into one hour. (The course is still being taught but someone else has taken over the wand.) Of course, the order in which material is taught need not be the same as the written account.

As someone once said, Mathematics is not a spectator sport. To learn and understand Mathematics you have to get stuck in and get your hands dirty. You have to do the calculations, manipulations, and proofs yourself, not just read the stuff and pretend you understand it. Thus I have included over 200 exercises to help with this process. I have also written a more or less complete set of solutions to these exercises. But these are not available in the book, for it is too easy simply to look up a solution. When you can't see how to do it you have to sweat a bit to find a solution. Someone else once said that horses sweat, gentlemen perspire, and ladies glow. However, I can't remember meeting many horses who could do mathematics all that well. In other words, although effort is important to learn mathematics you also need something else. You need help every now and then. That is why there are exercises *and* solutions. These solutions are available at

    www.cambridge.org/simmons

The book is divided into six Chapters, each chapter is divided into several Sections, and a few of these are divided into Blocks (Subsections). Each chapter contains a list of Items, that is Definitions, Lemmas, Theorems, Examples, and so on. These are numbered by section. Thus item $X.Y.Z$ is in Chapter X, Section Y, and is the Zth item in that section. Where a section is divided into blocks the items are still numbered by the parent section.

Each section contains a selection of Exercises. These are numbered separately throughout the section. Thus Exercise $X.Y.Z$ is in Chapter X, Section Y, and is the Zth exercise of that section. Again, where a section is divided into blocks the exercises are still numbered by the parent section.

Occasionally you will see a word or two IN THIS FONT. This is a mention of a NOTION that is dealt with in more detail later. You should remember to come back to this place when you understand the notion.

There are several other books available on this subject. Some of these are introductory texts and some are more advanced. I have listed some of them in the bibliography. None of these are needed when reading this book, but some will certainly help broaden and advance your understanding of the subject. I have refrained from passing comment on these books, for I know that different people have different tastes. However, you should look around for different accounts. Some of these will help.

I first became aware of Category Theory in 1965 during a Summer Meeting in Leicester (England). Since then I have been trying to learn and understand the subject. It is patently obvious to me that Category Theory is a very useful tool. It helps to organize many parts of mathematics. It can sort out the 'routine' aspects of a proof and isolate the 'essential content' of the result. In some ways that is why Eilenberg and MacLane invented the subject. However, I am not one of those 42ers who think that Category Theory is the essential foundations for Mathematics, Life, and Everything. Of course Category Theory is something that every mathematician should know something about, but there are other things as well.

Many people have influenced this book. For several years Andrea Schalk has used the material to teach an introductory course. Hugh Steele, Roman Krenický, and Francisco Lobo have pointed out and sometimes corrected my eccentricities. And Wolfy has guided me through some of the deeper mysteries of LaTeX. Where would we be without the wonderful LaTeX?

There may still be mistakes, inaccuracies, or garbled bits in the book. I would be quite happy to pass on the blame, but I won't. I am not a politician. I am responsible for everything inside the cover. The outside cover is the work of others.

Any book of this kind must contain many diagrams, some of which must commute. I have used Paul Taylor's diagram package to do this job. If you don't know this package then I recommend you have a look at it. I have also used his lesser known tree drawing package at one place.

At Cambridge University Press my contact, Silvia Barbina, has been very helpful. I once taught her a little bit about football (and, as she reminded me, some Model Theory). Silvia has made writing this final version very enjoyable. She has kept me on the straight and narrow, so I didn't wander off to do something else. In her charming Italian style she asked me (instructed me) to cut out all the jokes. This was quite difficult since some of the official categorical terminology is a joke, but I have done my best.

Clare Dennison and Lucy Edwards oversaw the production period (getting my raw code converted into the material you have in front of you). Siriol Jones copy-edited the book and corrected many of my silly mistakes. I thank them all. Roger Astley was chief pie-man for the whole project.

On a more personal level I am very grateful to Bobby Manc and what he is achieving. I hope he continues for quite some time. The Lodge (Appleby Lodge) is at last getting back to what it should be. Ruth Maddocks kept me cheerful. She made me the odd cup of tea. A very odd cup of tea.

Enjoy yourself and learn something at the same time.

1

Categories

This chapter gives the definition of 'category' in Section 1.1, and follows that by four sections devoted entirely to examples of categories of various kinds. If you have never met the notion of a category before, you should quite quickly read through Definition 1.1.1 and then go to Section 1.2. There you will find some examples of categories that you are familiar with, although you may not have recognized the categorical structure before. In this way you will begin to see what Definition 1.1.1 is getting at. After that you can move around the chapter as you like.

Remember that it is probably better not to start at this page and read each word, sentence, paragraph, ..., in turn. Move around a bit. If there is something you don't understand, or don't see the point of, then leave it for a while and come back to it later.

Life isn't linear, but written words are.

1.1 Categories defined

This section contains the definition of 'category', follows that with a few bits and pieces, and concludes with a discussion of some examples. No examples are looked at in detail, that is done in the remaining four sections. Section 1.2 contains a collection of simpler examples, some of which you will know already. You might want to dip into that section as you read this section. In the first instance you should find a couple of examples that you already know. As you become familiar with the categorical ideas you should look at some of the more complicated examples given in the later sections.

The following definition doesn't quite give all the required information. There are a couple of restrictions that are needed and which are described in detail in the paragraphs following.

1.1.1 Definition   A category C consists of

- a collection Obj of entities called **objects**
- a collection Arw of entities called **arrows**
- two assignments
$$Arw \xrightarrow[\text{target}]{\text{source}} Obj$$
- an assignment
$$Obj \xrightarrow{\text{id}} Arw$$
- a partial composition $Arw \times Arw \longrightarrow Arw$

where this data must satisfy certain restrictions as described below.   □

Before we look at the restrictions on this data let's fix some notation.

- We let $A, B, C, \ldots$ range over objects.
- We let $f, g, h, \ldots$ range over arrows.

This convention isn't always used. For instance, sometimes $a, b, c, \ldots$ range over objects, and $\alpha, \beta, \gamma, \ldots$ or $\theta, \phi, \psi, \ldots$ range over arrows. The notation used depends on what is convenient at the time and what is the custom in the topic under discussion. Here we will take the above convention as the norm, but sometimes we will use other notations.

There are two assignments

$$\text{source} \qquad \text{target}$$

each of which attaches an object to an arrow, that is each consumes an arrow and returns an object. We write

$$A \xrightarrow{\ f\ } B$$

to indicate that f is an arrow with source A and target B. This is a small example of a **diagram**. Later we will see some slightly bigger ones.

This terminology isn't always used. Sometimes combinations of

$$A \xrightarrow{\hspace{3cm} f \hspace{3cm}} B$$

source	arrow	target
domain	morphism	codomain
	map	

are used. Certainly morphisms (such as group morphisms) and maps (such as continuous maps) usually are examples of arrows in some category. However, it is better to use 'arrow' for the abstract notion, and so distinguish between the general and the particular.

The word 'domain' already has other meanings in mathematics. Why bother with this and 'codomain' when there are two perfectly good words that capture the idea quite neatly. You will also see

$$f : A \longrightarrow B$$

used to name the arrow above. However, as we see later, you should not think of an arrow as a function.

All three of the notations

$$A \xrightarrow{\quad id_A \quad id_A \quad 1_A \quad} A$$

are used for the identity arrow assigned to the object A. We will tend to use id_A. Notice that the source and the target of id_A are both the parent object A. Quite often when there is not much danger of confusion id is written for id_A. You will also find in the literature that some people write 'A' for the arrow id_A. This is a notation so ridiculous that it should be laughed at in the street.

Certain pairs of arrows are compatible for composition to form another arrow. Two arrows

$$A \xrightarrow{\quad f \quad} B_1 \qquad\qquad B_2 \xrightarrow{\quad g \quad} C$$

are composible, in that order, precisely when B_1 and B_2 are the same object, and then an arrow

$$A \longrightarrow C$$

is formed. For arrows

$$A \xrightarrow{\quad f \quad} B \xrightarrow{\quad g \quad} C$$

both of the notations

$$A \xrightarrow{\quad g \circ f \quad gf \quad} C$$

are used for the composite arrow. Read this as

$$g \text{ after } f$$

and be careful with the order of composition. Here we write $g \circ f$ for the composite.

We need to understand how to manipulate composition, sometimes involving many arrows.

Composition of arrows is associative as far as it can be. For arrows

$$A \xrightarrow{\ f\ } B \xrightarrow{\ g\ } C \xrightarrow{\ h\ } D$$

various composites are possible, as follows.

$$A \xrightarrow{\qquad\qquad (h \circ g) \circ f \qquad\qquad} D$$
$$A \xrightarrow{\ f\ } B \xrightarrow{\qquad h \circ g \qquad} D$$
$$A \xrightarrow{\ f\ } B \xrightarrow{\ g\ } C \xrightarrow{\ h\ } D$$
$$A \xrightarrow{\qquad g \circ f \qquad} C \xrightarrow{\ h\ } D$$
$$A \xrightarrow[\ h \circ (g \circ f)\ ]{\qquad\qquad\qquad} D$$

It is required that the two extreme arrows are equal

$$(h \circ g) \circ f = h \circ (g \circ f)$$

and we usually write

$$h \circ g \circ f$$

for this composite. This is the first of the axioms restricting the data.

The second axiom says that identity arrows are just that. Consider

$$A \xrightarrow{\ id_A\ } A \xrightarrow{\ f\ } B \xrightarrow{\ id_B\ } B$$

an arbitrary arrow and the two compatible identity arrows. Then

$$id_B \circ f = f = f \circ id_A$$

must hold.

Given two objects A and B in an arbitrary category C, there may be no arrows from A to B, or there may be many. We write

$$C[A, B] \quad \text{or} \quad C(A, B)$$

for the collection of all such arrows. For historical reasons this is usually called the

hom-set

from A to B, although

arrow-class

would be better. Some people insist that $C[A, B]$ should be a set, not a class. As usual, there are some variants of this notation. We often write

$$[A, B] \quad \text{for} \quad C[A, B]$$

especially when it is clear which category C is intended. Sometimes

$$\mathrm{Hom}_C[A, B]$$

is used for this hom-set.

We have seen above one very small diagram. Composition gives us a slightly larger one. Consider three arrows

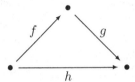

arranged in a triangle, as shown. Here we haven't given each object a name, because we don't need to. However, the notation does *not* mean that the three objects are the same. For this small diagram, the triangle, the composite $g \circ f$ exists to give us a **parallel pair**

of arrows across the bottom of the triangle. These two arrows may or may not be the same. When they are

$$h = g \circ f$$

we say the triangle **commutes**. We look at some more commuting diagrams in Section 2.1, and other examples occur throughout the book.

Examples of categories

In the remaining sections of this chapter we look at a selection of examples of categories. Roughly speaking these are of four kinds.

The first collection is listed in Table 1.1 on page 6. These all have a similar nature and are examples of the most common kind of category we meet in practice. In each an object is a structured set, a set furnished, or equipped, with some extra gadgetry, the **furnishings** of the object. An arrow between two objects is a function between the carrying sets where the function 'respects' the carried structure. Arrow composition is then function composition. We look at some of these categories in Section 1.2.

Some categories listed in Table 1.1 are not defined in this chapter. Some are used later to illustrate various aspects of category theory, in which case each

Table 1.1 *Categories of structured sets and structure preserving functions*

Category	Objects	Arrows
Set	sets	total functions
Pfn	sets	partial functions
Set$_\perp$	pointed sets	point preserving functions
RelH	sets with a relation	relation respecting functions
Sgp	semigroups	morphisms
Mon	monoids	morphisms
CMon	commutative monoids	morphisms
Grp	groups	morphisms
AGrp	abelian groups	morphisms
Rng	rings	morphisms
CRng	commutative rings	morphisms
Pre	pre-ordered sets	monotone maps
Pos	posets	monotone maps
Sup	complete posets	$\bigvee$-preserving monotone functions
Join	posets with all finitary joins	$\vee$-preserving monotone functions
Inf	complete posets	$\bigwedge$-preserving monotone functions
Meet	posets with all finitary meets	$\wedge$-preserving monotone functions
Top	topological spaces	continuous maps
Top$_\star$	pointed topological spaces	point preserving continuous maps
Top^{open}	topological spaces	continuous open maps
Vect$_K$	vector spaces over a given field K	linear transformations
Set-R	sets with a right action from a given monoid R	action preserving functions
R-**Set**	sets with a left action from a given monoid R	action preserving functions
Mod-R	right R-modules over a ring R	morphisms
R-**Mod**	left R-modules over a ring R	morphisms

Table 1.2 *More complicated categories*

Category	Objects	Arrows
$\boldsymbol{RelA}$	sets	binary relations
$\boldsymbol{Pos}^{\dashv}$	posets	poset adjunctions
$\boldsymbol{Pos}^{pp}$	posets	projection embedding pairs
$\widehat{S}$	presheaves on a given poset S	natural transformations
$\widehat{C}$	presheaves on a given category C	natural transformations
$\boldsymbol{Ch}(\boldsymbol{Mod}\text{-}R)$	chain complexes	

is defined when it first appears. Some categories are listed but not used in this book, but you should be able to fill in the details when you need to.

These simple examples tend to give the impression that in any category an object is a structured set and an arrow is a function of a certain kind. This is a false impression, and in Section 1.3 we look at some examples to illustrate this. In particular, these examples show that an arrow need not be a function (of the kind you first thought of).

An important message of category theory is that the more important part of a category is not its objects but the way these are compared, its arrows. Given this we might expect a category to be named after its arrows. For historical reasons this often doesn't happen.

Section 1.4 contains some examples to show that the objects of a category can have a rather complicated internal structure, and the arrows are just as complicated. These examples are important in various parts of mathematics, but you shouldn't worry if you cannot understand them immediately.

Table 1.2 lists some of these more complicated examples looked at in Sections 1.3 and 1.4.

Finally in Section 1.5 we look at two very simple kinds of categories. These examples could be given now, but in some ways it is better if we leave them for a while.

Exercises

1.1.1 Observe that sets and functions do form a category $\boldsymbol{Set}$.

1.1.2 Can you see that each poset is a category, and each monoid is a category? Read that again.

1.2 Categories of structured sets

The categories we first meet usually have a rather simple nature. Each object is a structured set

$$(A, \cdots)$$

a set furnished with some extra gadgetry, its furnishings, and each arrow

$$(A, \cdots) \longrightarrow (B, \cdots)$$

is a (total) function

$$f : A \longrightarrow B$$

between the two carrying sets which respects the carried structure in some appropriate sense. More often than not these structured sets are 'algebras'. Thus the furnishings carried by A are a selection of nominated elements, and a selection of nominated operations on A. These operations are usually binary or singulary, but other arities do occur.

You have already met

$$\textbf{Grp} \quad \textbf{Rng} \quad \textbf{Vect}_K$$

as given in Table 1.1, but you may not have realized that each of these is a category. You should make sure that you understand the workings of each of these as a category of 'algebras'. You may have to puzzle a bit over $\textbf{Vect}_K$, but later we look at some more general examples of this nature, and that should help you.

To help with the general idea, in the first part of this section we look at the category $\textbf{Mon}$ of monoids. This has all the typical properties of an 'algebraic' category. You may not have met monoids before, so this example will serve as an introduction, and it is quite easy to understand. Monoids are quite important in category theory. They can tell us quite a lot about the structure of a particular category. Also, they can be used to illustrate many aspects of category theory.

The exercises for the first part of this section look at several other categories of structured sets, some of which are not 'algebraic' in this intuitive sense. One of these

$$\textbf{Top}$$

is particularly important, and you should make sure you understand it. It is important here and in many other parts of mathematics.

1.2.1 Example A monoid is a structure

$$(R, \star, 1)$$

where R is a set, $\star$ is a binary operation on R (usually written as an infix), 1 is a nominated element of R, and where

$$(r \star s) \star t = r \star (s \star t) \qquad 1 \star r = r = r \star 1$$

for all $r, s, t \in R$. In other words, the operation is associative and the nominated element is a unit for the operation. Monoids are sometimes referred to as unital semigroups, or even semigroups. However, sometimes a 'semigroup' need not have a unit.

Usually we omit the operation symbol and write

$$rs \quad \text{for} \quad r \star s$$

but for the time being we will stick to the official notation.

A monoid morphism

$$R \xrightarrow{\phi} S$$

between two monoids is a function that respects the furnishings, that is

$$\phi(r \star s) = \phi(r) \star \phi(s) \qquad \phi(1) = 1$$

for all $r, s \in R$. (Notice that we have overloaded the operation symbol and the unit symbol. That shouldn't cause a problem here, but every now and then it is a good idea to distinguish between the source and target furnishings.)

It is routine to check that for two morphisms

$$R \xrightarrow{\phi} S \xrightarrow{\psi} T$$

between monoids the function composite

$$R \xrightarrow{\psi \circ \phi} T$$

is a morphism.

This gives us the category **Mon** of monoids (as objects) and monoid morphisms (as arrows). The verification of the axioms is almost trivial. Given a monoid R the identity arrow

$$R \xrightarrow{id_R} R$$

is just the identity function on R viewed as a morphism. □

As suggested above, many categories fit into this 'algebraic' form. Each object is a structured set, and each arrow (usually called a morphism or a map) is a structure respecting function. Almost all of the categories in Table 1.1 fit into this kind, but one or two don't.

In a sense the study of monoids is the study of composition in the miniature. There is a corresponding study of comparison in the miniature. That is the topic of the next example.

1.2.2 Example   A pre-order $\leq$ on a set S is a binary relation that is both re-flexive and transitive. (Sometimes a pre-order is called a quasi-order.) A partial order is a pre-order that is also anti-symmetric.

A

preset   poset

is a set S furnished with a

pre-order   partial order

respectively. Thus each poset is a preset, but not conversely.

When comparing two such structures

$$(R, \leq_R) \qquad (S, \leq_S)$$

we use the carrying sets R and S to refer to the structures and write $\leq$ for both the carried comparisons. Rarely does this cause any confusion, but when it does we are a bit more careful with the notation.

Given a pair R, S of presets a **monotone map**

$$R \xrightarrow{\ f\ } S$$

is a function, as indicated, such that

$$x \leq y \implies f(x) \leq f(y)$$

for all $x, y \in R$. Note that this condition is an implication, not an equivalence. It is routine to check that for two monotone maps

$$R \xrightarrow{\ f\ } S \xrightarrow{\ g\ } T$$

between presets the function composition $g \circ f$ is also monotone.

This gives us two categories

Pre   ***Pos***

where the objects are

presets   posets

respectively, and in both cases the arrows are the monotone maps. Each identity arrow is the corresponding identity function viewed as a monotone map. $\square$

Consider a pair R and S of posets. Each is a preset, so we have the two collections of arrows

$$\boldsymbol{Pre}[R,S] \qquad \boldsymbol{Pos}[R,S]$$

in the categories. A moment's thought shows that, as sets of functions, these two sets are the same. Technically, this shows that $\boldsymbol{Pos}$ is a FULL SUBCATE-GORY of $\boldsymbol{Pre}$.

The study of monoids is the study of composition in the miniature.

The study of presets is the study of comparison in the miniature.

What should we do to study these two notions together and in the large? Category theory! In a sense every category is an amalgam of certain monoids and presets, and that is a good enough reason why we should always keep these two simple notions in mind.

From the examples we have seen so far it is easy to get the impression that certain things always happen. The next example shows that some categories can be awkward (and sometimes cantankerous).

1.2.3 Example   We enlarge the category $\boldsymbol{Set}$ of sets and total functions to the category $\boldsymbol{Pfn}$ of sets and partial functions. The objects of $\boldsymbol{Pfn}$ are just sets

$$A, B, C, \ldots$$

as in $\boldsymbol{Set}$. However, an arrow

$$A \xrightarrow{\quad f \quad} B$$

is a *partial* function from A to B. In other words, an arrow is a total function

from a subset X of the source A. (This is an example where the use of the word 'domain' for source can be confusing. The set X is the domain of definition of the partial function.) Notice that we need to distinguish between the total function $\overline{f}$ and the arrow f it determines. The notation has been chosen to emphasize that distinction.

We wish to show that these objects and arrows form a category $\boldsymbol{Pfn}$. To do that we must first produce a composition of arrows.

Consider a pair of partial functions.

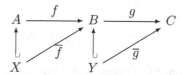

How might we compose these? We somehow want to stick $\overline{f}$ and $\overline{g}$ together, but these functions are not composition compatible.

We extract a subset $U \subseteq A$ by

$$a \in U \iff a \in X \text{ and } \overline{f}(a) \in Y$$

(for $a \in A$). Since $\overline{f}$ is defined on the whole of U we restrict $\overline{f}$ to U.

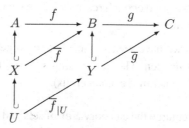

Now we do have composition compatible functions. Thus we take

$$g \circ f$$

to be that arrow (partial function) determined by

$$\overline{g \circ f} = \overline{g} \circ \overline{f}_{|U}$$

to produce a composition of arrows in *Pfn*.

Notice here how the symbol '∘' is overloaded. On the right it is the standard composition of total functions. On the left it is the defined operation on partial functions. If at first you find this confusing then write '•' for the defined operation. Thus

$$\overline{g \bullet f} = \overline{g} \circ \overline{f}_{|U}$$

is its definition.

There is still some work to be done. For instance, we need to show that this composition of arrows is associative. That is left as an exercise. □

Once we see it the step from *Set* to *Pfn* is not so big. An arrow is still a function, but we have to take a little more care with composition. There is also a much neater way of handling *Pfn*. Perhaps you can think about that.

We began this section by looking at the category **Mon** of monoids. We conclude by looking at two categories attached to each monoid.

1.2.4 Example   Let R be a fixed, but arbitrary, monoid. A

left   right

R-set is a set A together with an action

$$R, A \longrightarrow A \qquad\qquad A, R \longrightarrow A$$
$$r, a \longmapsto ra \qquad\qquad a, r \longmapsto ar$$

where

$$s(ra) = (sr)a \qquad\qquad (ar)s = a(rs)$$
$$1a = a \qquad\qquad a = a1$$

for each $a \in A$ and $r, s \in R$. Here the two definitions are given in parallel. These R-sets are the objects of two categories

R-**Set**   **Set**-R

with left R-sets on the left and right R-sets on the right.

Given two R-sets A and B of the same handedness, a morphism

$$A \xrightarrow{\quad f \quad} B$$

is a function f such that

$$f(ra) = rf(a) \qquad\qquad f(ar) = f(a)r$$

for each $a \in A$ and $r \in R$. These are the arrows of the two categories.   □

This may look a quite simple example but it is useful. Many aspects of category theory can be illustrated with these categories. We use them quite a lot in this book. They are also module categories in miniature. We can replace the monoid R by a ring and replace each set A by an abelian group. This gives the categories

R-**Mod**   **Mod**-R

of left and right modules over R, respectively. These categories have quite a bit more structure, but we won't go into that too much here.

Exercises

1.2.1   The category **Pno** described in this exercise may look less than exciting, but it plays an important role in mathematics. (It was originally discovered
by Dedekind without the category theory.)

The objects of **Pno** are the structures (A, α, a) where A is a set, and where
$\alpha : A \longrightarrow A$ is a function, and $a \in A$ is a nominated element. Given two
such structures a morphism

$$(A, \alpha, a) \xrightarrow{\ f\ } (B, \beta, b)$$

is a function $f : A \longrightarrow B$ which preserves the structure in the sense that

$$f \circ \alpha = \beta \circ f \qquad f(a) = b$$

hold.

(a) Verify that **Pno** is a category.

(b) Show that $(\mathbb{N}, \mathrm{succ}, 0)$ is a **Pno**-object (where succ is the successor
function).

(c) Show that for each **Pno**-object (A, α, a) there is a unique arrow

$$(\mathbb{N}, \mathrm{succ}, 0) \longrightarrow (A, \alpha, a)$$

and describe the behaviour of the carrying function.

1.2.2   Consider pairs (A, X) where A is a set and $X \subseteq A$. For two such pairs
a morphism

$$(A, X) \xrightarrow{\ f\ } (B, Y)$$

is a function $f : A \longrightarrow B$ that respects the selected subsets, that is

$$f(x) \in Y$$

for each $x \in X$. Show that such pairs and morphisms form a category **SetD**,
the category of sets with a distinguished subset.

1.2.3   Consider pairs (A, R) where A is a set and $R \subseteq A \times A$ is a binary
relation on A. Show that these pairs are the objects of a category. You must
find a sensible notion of morphism for such pairs.

1.2.4   A topological space $(S, \mathcal{O}S)$ is a set S furnished with a certain family
$\mathcal{O}S$ of subsets of S (called the open sets of the space). This family is required
to contain both $\emptyset$ and S, be closed under $\cap$ (binary intersection), and be closed
under $\bigcup$ (arbitrary unions).

A continuous map

$$(S, \mathcal{O}S) \xrightarrow{\phi} (T, \mathcal{O}T)$$

between two such spaces is a function

$$\phi : S \longrightarrow T$$

such that

$$\phi^{\leftarrow}(V) \in \mathcal{O}S$$

for each $V \in \mathcal{O}T$. Here $\phi^{\leftarrow}$ is the inverse image map given by

$$x \in \phi^{\leftarrow}(V) \Longleftrightarrow \phi(x) \in V$$

for each $V \in \mathcal{O}T$ and $x \in S$.

Show that these spaces and maps form a category **Top**.

1.2.5 Let A be an arbitrary object of an arbitrary category C. Show that $C[A, A]$ is a monoid under composition.

1.2.6 Fill in the details missing from the description of **Pfn**. In particular, you should show that composition of partial functions is associative.

1.2.7 A pointed set is a set S with a nominated element which we usually write as $\perp$. An arrow

$$S \xrightarrow{\phi} T$$

between two such pointed sets is a function $\phi : S \longrightarrow T$ which respects the nominated points, that is $\phi(\perp) = \perp$.

Almost trivially, pointed sets with these arrows form a category **Set$_\perp$**.

Try to show that **Set$_\perp$** and **Pfn** are 'essentially the same' category.

1.2.8 Verify that for each monoid R both

$$R\text{-}\textbf{Set} \qquad \textbf{Set-}R$$

are categories.

Can you see how each is a category of structured sets?

1.3 An arrow need not be a function

In this section we look first at some examples to show that arrows may not be the simple kind of things we have seen so far. Then we look at some general constructions for turning an old category into a new one.

In the next example an arrow is still a function, but not where you might expect it to be.

1.3.1 Example   The objects are the finite sets. An arrow

$$A \xrightarrow{\ f\ } B$$

is a function

$$f : A \times B \longrightarrow \mathbb{R}$$

(with no imposed conditions). For each pair

$$A \xrightarrow{\ f\ } B \xrightarrow{\ g\ } C$$

of arrows we define

$$g \circ f : A \times C \longrightarrow \mathbb{R}$$

by

$$(g \circ f)(a, c) = \sum \left\{ f(a, y)g(y, c) \,|\, y \in B \right\}$$

for $a \in A, b \in B$. A little work shows that this produces a category.   □

We did not give this category a name because it is not that important. It is merely an example to illustrate that an arrow need not be a function in the way you might expect it to be.

1.3.2 Example   We have seen that the category *Set* of sets and functions can be extended to *Pfn* by adding more arrows but keeping the same objects. There is also a different extension to the category *RelA*.

Again the objects of *RelA* are just sets. However, a *RelA*-arrow

$$A \xrightarrow{\ F\ } B$$

is a subset $F \subseteq B \times A$ which we can think of as a relation from A to B. In other words, *RelA*$[A, B]$ is just the set of all such relations from A to B. You should note the way the source and target have been set up. This is not a mistake. It leads to a neater description of the category.

Before we can claim this is a category we must define the composition of these arrows, and then check that the axioms are satisfied.

Consider an arrow F as above, so $F \subseteq B \times A$. For $a \in A$ and $b \in B$ we write bFa for $(b, a) \in F$. For two composible arrows

$$A \xrightarrow{\ F\ } B \xrightarrow{\ G\ } C$$

we define the composition $G \circ F$ by

$$c(G \circ F)a \Longleftrightarrow (\exists b \in B)[cGbFa]$$

for $a \in A, b \in B$. We show that a is $G \circ F$ related to c by passing through a common element $b \in B$. It is easy to check that this composition is associative, and the equality relation on a set gives the identity arrow.

The two categories **Set** and **RelA** are connected in a certain way (which will be explained in more detail later). There is a canonical way

$$A \xrightarrow{\ f\ } B \qquad \longmapsto \qquad A \xrightarrow{\ \Gamma(f)\ } B$$

of converting a **Set**-arrow into a **RelA**-arrow with the same source and target. We simply take the graph of the function, that is we let

$$b\,\Gamma(f)\,a \Longleftrightarrow b = f(a)$$

for $a \in A, b \in B$. $\qquad\square$

The next example is important in itself, and also provides a miniature version of a central notion of category theory, that of an ADJUNCTION.

1.3.3 Example   We modify the category **Pos** of posets, of Example 1.2.2, to produce a new category $\textbf{\textit{Pos}}^{\dashv}$. As with **Pos**, the objects of $\textbf{\textit{Pos}}^{\dashv}$ are posets, but the arrows are different.

Given a pair S, T of posets, an adjunction from S to T is a pair of monotone maps as on the left such that the equivalence on the right

$$S \underset{g}{\overset{f}{\rightleftarrows}} T \qquad f(a) \le b \Longleftrightarrow a \le g(b)$$

holds for all $a \in S$ and $b \in T$. We call

$$f \text{ the left adjoint} \qquad g \text{ the right adjoint}$$

of the pair, and sometimes write

$$f \dashv g$$

to indicate an adjunction.

Here we use the more common notation and write

$$S \xrightarrow[\quad f_* \quad]{\quad f^* \quad} T$$

to indicate an adjunction $f^* \dashv f_*$. Sometimes a harpoon arrow

$$S \xrightarrow{\quad f^* \dashv f_* \quad} T$$

is used to indicate an adjunction. By convention, an adjunction points in the direction of its left component. Thus S is the source and T is the target. (You are warned that in some of the older literature this convention hadn't yet been established.)

Poset adjunctions are the arrows of $\boldsymbol{Pos}^{\dashv}$.

This gives us the object and arrows of $\boldsymbol{Pos}^{\dashv}$, but we still have some work to do before we know we have a category.

Consider a pair of adjunctions.

$$R \xrightarrow{\quad f^* \dashv f_* \quad} S \xrightarrow{\quad g^* \dashv g_* \quad} T$$

which ought to be composible. How should the composite

$$R \xrightarrow{\quad (g^* \dashv g_*) \circ (f^* \dashv f_*) \quad} T$$

be formed? The two left hand components are monotone maps that compose to give a monotone map. Similarly the two right hand components are monotone maps that compose to give a monotone map. Thus we have a pair of monotone maps

$$R \xrightarrow[\quad f_* \circ g_* \quad]{\quad g^* \circ f^* \quad} T$$

going in opposite directions. We check that this is an adjunction and take that as the composite. Almost trivially, this composition is associative, and so we do obtain a category. $\square$

It is not so surprising that any given monotone map may or may not have a left adjoint, and it may or may not have a right adjoint. It can have neither, and it can have one without the other. What is a little surprising is that it can have both adjoints where these are not the same. In fact, arbitrarily longs strings of adjoints can be produced. A simple example of this is given in Chapter 6.

Once we become familiar with categories we find that old categories can be used to produce new categories. Let's look at some examples.

1.3.4 Example Consider categories C and D. To help us distinguish between these let us write

$A, B, C \ldots$ for objects of C $f, g, h \ldots$ for arrows of C

$R, S, T \ldots$ for objects of D $\theta, \phi, \psi \ldots$ for arrows of D

respectively. We form a new category, the product

$$C \times D$$

of C and D as follows. Each new object is an ordered pair of old objects

$$(A, R)$$

an object A from C and an object R from D. A new arrow

$$(A, R) \longrightarrow (B, S)$$

is a pair of old arrows

$$A \xrightarrow{\;\;f\;\;} B \qquad\qquad R \xrightarrow{\;\;\theta\;\;} S$$

from the given categories. For composible new arrows

$$(A, R) \xrightarrow{\;(f, \theta)\;} (B, S) \xrightarrow{\;(g, \phi)\;} (C, T)$$

the composite

$$(A, R) \xrightarrow{\;(g \circ f, \phi \circ \theta)\;} (C, T)$$

is formed using composition in the old categories. Almost trivially, this does give a category. $\qquad\qquad\qquad\qquad\qquad\qquad\qquad\qquad\qquad\qquad\qquad\square$

That's not the most exciting example you have ever seen, is it? Here is a more interesting construction.

1.3.5 Example Given a category C we form a new category where the new objects are the arrows of C. This is the arrow category of C.

Consider the small graph

$$
\begin{array}{c}
0 \\
(\downarrow) \qquad\qquad \Big\downarrow \\
1
\end{array}
$$

with two nodes, here labelled 0 and 1, and with one edge. We use $(\downarrow)$ to convert C into a new category

$$C^{\downarrow}$$

the category of $(\downarrow)$-diagrams in C.

We think of ($\downarrow$) as a TEMPLATE for diagrams in C, and these diagrams are the objects of $C^{\downarrow}$. Thus a new object is a pair of old objects

$$
\begin{array}{c}
A_0 \\
| \\
\alpha \\
\downarrow \\
A_1
\end{array}
$$

and an old arrow between them. Given two new objects a new arrow

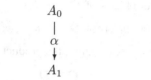

is a pair of old arrows

$$
\begin{array}{ccc}
A_0 & \xrightarrow{\ f_0\ } & B_0 \\
\downarrow{\alpha} & & \downarrow{\beta} \qquad f_1 \circ \alpha = \beta \circ f_0 \\
A_1 & \xrightarrow[\ f_1\ ]{} & B_1
\end{array}
$$

such that the square commutes. Composition of new arrows is performed in the obvious way, we compose the two component old arrows. You should check that this does give a category.   □

This is a simple example of a much more general construction, that of a FUNCTOR CATEGORY. We look at this once we know what a FUNCTOR is. Other simple examples of this construction are given in the exercises.

The idea of the previous example is to view *all* the arrows of the old category as the objects of the new category. Sometimes we want to do a similar thing but using only *some* old arrows.

1.3.6 Example   Given a category C and an object S we form the two slice categories

$$
(C \downarrow S) \qquad\qquad (S \downarrow C)
$$

of objects

over S \qquad\qquad under S

respectively. Each object of the new category is an arrow

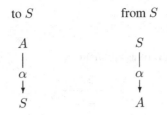

to S from S

of $\mathcal{C}$. An arrow of the new category

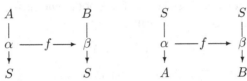

is an arrow of $\mathcal{C}$

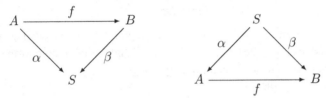

for which the indicated triangle commutes. Composition of the new arrows is obtained from composition of arrows in $\mathcal{C}$ □

As with Example 1.3.5 this construction is a particular case of a more general construction, that of a COMMA CATEGORY. Before we can explain that we need to understand the notion of a FUNCTOR.

Exercises

1.3.1 Consider the strictly positive integers $1, 2, 3, \ldots$ as objects. For two such integers m, n let an arrow

$$n \longrightarrow m$$

be an $m \times n$ matrix A (with real entries). Given two compatible matrices

$$n \xrightarrow{\;\;B\;\;} k \qquad k \xrightarrow{\;\;A\;\;} m$$

let the composite

$$n \xrightarrow{\;\;A \circ B\;\;} m$$

be the matrix product AB. Show that this gives a category.

Can you show that this example is a bit of a cheat?

1.3.2 A **directed graph**, or simply a **graph** for short (and sometimes called a network), is a pair

$$(N, E)$$

of sets together with a pair of assignments

$$E \underset{\tau}{\overset{\sigma}{\longrightarrow}} V$$

(as with a category). Each member of N is a **node**, and each member of E is an **edge**. For each edge $e \in E$ we call

$$\sigma(e) \qquad \tau(e)$$

the source node and the target node of e, and we write

$$a \overset{e}{\longrightarrow} b$$

to indicate that $\sigma(e) = a$ and $\tau(e) = b$. In general there are no other conditions on these edges and nodes. In particular, there is no notion of composing edges. Notice that (modulo size) each category is a graph.

A **graph morphism**

$$(N, E) \overset{f}{\longrightarrow} (M, F)$$

is a pair of functions

$$N \overset{f_0}{\longrightarrow} M \qquad\qquad E \overset{f_1}{\longrightarrow} F$$

such that

$$\sigma \circ f_1 = f_0 \circ \sigma \qquad \tau \circ f_1 = f_0 \circ \tau$$

hold. Of course, here there are two different source and two different target assignments.

Show that, with the appropriate notion of composition, graphs and their morphisms form a category.

1.3.3 Consider any pair of categories A and S. We form a new category. The new objects are pairs

$$(A, R)$$

where A is an A-object and R is an S-object. A new arrow

$$(A, R) \overset{(f, \phi)}{\longrightarrow} (B, S)$$

is a pair of arrows

$$A \xrightarrow{\quad f \quad} B \qquad\qquad R \xleftarrow{\quad \phi \quad} S$$

from the two component categories where the S-arrow goes backwards.

Show that with the obvious composition this does form a category.

1.3.4   As in Exercise 1.2.8, each monoid R gives us a category *Set-R* of (right) R-sets. We can also vary R to produce a larger category.

Each object is a pair

$$(A, R)$$

where R is a monoid and A is an R-set. Each arrow

$$(A, R) \xrightarrow{\quad (f, \phi) \quad} (B, S)$$

is a pair

$$A \xrightarrow{\quad f \quad} B \qquad\qquad R \xleftarrow{\quad \phi \quad} S$$

where ϕ is a monoid morphism and f is a function with

$$f(a\phi(s)) = f(a)s$$

for each $a \in A$ and $s \in S$.

Using the obvious composition, show that this does give a category.

1.3.5   Consider the category *RelA* of Example 1.3.2.

Show the defined composition is associative, and so it is a category. Show also that

$$\Gamma(g \circ f) = \Gamma(g) \circ \Gamma(f)$$

for each pair of composible *Set*-arrows.

1.3.6   Consider any pair of *Pos*-arrows.

$$S \underset{g}{\overset{f}{\rightleftarrows}} T$$

(a) Show that $f \dashv g$ precisely when both $id_S \leq g \circ f$ and $f \circ g \leq id_T$, where the two comparisons are pointwise.

(b) Show that if $f \dashv g$ then

$$f \circ g \circ f = f \qquad g \circ f \circ g = g$$

and hence $g \circ f$ is a closure operation on A and $f \circ g$ is a co-closure operation on B.

1.3.7   Posets and certain adjoint pairs form another category $\boldsymbol{Pos}^{pp}$.
The objects of $\boldsymbol{Pos}^{pp}$ are again just posets. A $\boldsymbol{Pos}^{pp}$-arrow

$$A \xrightarrow{\quad (f,g) \quad} B$$

is a $\boldsymbol{Pos}^{\dashv}$-arrow

$$A \xrightarrow{\quad f \dashv g \quad} B$$

for which $g \circ f = id_A$. These arrows are sometimes called projection pairs.

Show that these projection pairs are closed under composition, and hence
$\boldsymbol{Pos}^{pp}$ is a category.

You see here a useful little trick. It can be helpful to draw arrows in different,
but related, categories in a different way. Thus here we have

$$
\begin{array}{ll}
\boldsymbol{Pos} & \longrightarrow \\
\boldsymbol{Pos}^{\dashv} & \longrightarrow \\
\boldsymbol{Pos}^{pp} & \Longrightarrow
\end{array}
$$

for the three different kinds of arrows.

1.3.8   Consider the ordered sets $\mathbb{Z}$ and $\mathbb{R}$ as posets, and let

$$\mathbb{Z} \xrightarrow{\quad \iota \quad} \mathbb{R}$$

be the insertion.

(a) Show there are (unique) maps

$$\mathbb{R} \underset{\rho}{\overset{\lambda}{\rightrightarrows}} \mathbb{Z}$$

such that

$$\mathbb{Z} \xrightarrow{\quad \iota \dashv \rho \quad} \mathbb{R} \xrightarrow{\quad \lambda \dashv \iota \quad} \mathbb{Z}$$

are adjunctions.

(b) Show also that this composite is $id_{\mathbb{Z}}$ in $\boldsymbol{Pos}^{\dashv}$ and the other composite,
on $\mathbb{R}$, is idempotent.

(c) Show that $\iota \dashv \rho$ is a $\boldsymbol{Pos}^{pp}$-arrow, but $\lambda \dashv \iota$ is not.

1.3.9   For a poset S let $\mathcal{L}S$ be the poset of lower sections under inclusion. (A
lower section of S is a subset $X \subseteq S$ such that

$$y \le x \in X \implies y \in X$$

for all $x, y \in S$.)

(a) For a monotone map

$$T \xrightarrow{\phi} S$$

between posets, show that setting $f = \phi^{\leftarrow}$ (the inverse image map) produces a monotone map

$$\mathcal{L}T \xleftarrow{f = \phi^{\leftarrow}} \mathcal{L}S$$

in the opposite direction.

(b) Show that f has both a left adjoint and a right adjoint

$$f^{\sharp} \dashv f \dashv f_{\flat}$$

where, in general, these are different.

1.3.10 Let C be an arbitrary category. In Example 1.3.5 we used ($\downarrow$) as a template to obtain a category $C^{\downarrow}$ of certain diagrams from C. The same idea can be used with other templates.

A wedge in a category C is a pair or arrows

as shown. A wedge morphism

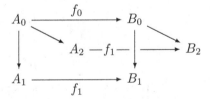

is a triple of arrows which make the two associated squares commute.

(a) Show that wedges and wedge morphisms form a category.

(b) This wedge example uses

as the template. Play around with other templates to produce other examples of categories. For example, consider each of

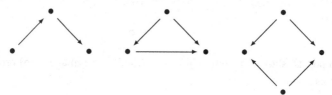

and worry about which cells are required to commute.

1.3.11   Let **1** and **2** be the 1-element set and the 2-element set, respectively. Describe the categories

$$(Set \downarrow 1) \quad (1 \downarrow Set) \quad (Set \downarrow 2) \quad (2 \downarrow Set)$$

and show that you have met two of them already together with near relatives of the other two.

1.3.12   Given a category C and two objects S, T we form

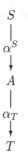

$(S \downarrow C \downarrow T)$

the **butty category** between S and T. Each object of the new category is an object A of C together with a pair of arrows from S and to T. An arrow of the new category is an arrow f of C to make the two triangles commute.

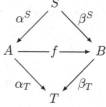

(a) Show that with the appropriate notion of composition this gives a category.

(b) Can you show that for an appropriate parent category C both the slice categories

$$(C \downarrow T) \qquad (S \downarrow C)$$

are instance of the butty construction?

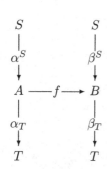

1.4 More complicated categories

From the examples we have seen so far you might conclude that category theory is making a bit of a fuss. It is true that objects are not just structured sets and arrows are not just functions, but the examples seem to suggest that we don't move too far from those ideas. Of course, as yet we have seen only comparatively simple examples of categories. One of the original aims of category theory was to organize and analyse what we now see as rather complicated categories. The simpler examples came along later. In this section we look at a couple of examples of the more complicated kind of category. You probably won't understand these at a first reading, but you should give it a go. You should come back to these examples as you learn more about category theory.

1.4.1 Example Let S be any partially ordered set. We describe the category $\widehat{S}$ of PRESHEAVES ON S. There is a more general notion where S is replaced by an arbitrary category, but we save that for later. We may think of $\widehat{S}$ as the category of 'sets developing over S'. At first sight the structure of $\widehat{S}$ looks quite complicated, but you will get used to it.

We think of S as a store of indexes $i, j, k, \ldots$ partially ordered

$$j \leq i$$

to form a poset.

A **presheaf** on S is an S-indexed family of sets

$$\text{A} \qquad \big(A(i) \mid i \in S\big)$$

together with a family of connecting functions

$$\mathcal{A} \qquad A(i) \xrightarrow{\ A(j,i)\ } A(j)$$

one for each comparison $j \leq i$. Note these functions progress *down* the poset. These functions have to fit together in a coherent fashion. Thus

$$A(i,i) = id_{A(i)}$$

for each index $i \in S$, and the triangle

$$A(k,j) \circ A(j,i) = A(k,i)$$

commutes for all $k \leq j \leq i$. These are the objects of $\widehat{S}$. Note the way the connecting functions are indexed.

An arrow

$$A \xrightarrow{\quad f \quad} B$$

between two presheaves is an S-indexed family of functions

$$A(i) \xrightarrow{\quad f_i \quad} B(i)$$

such that

$$
\begin{array}{ccc}
A(i) & \xrightarrow{\ f_i\ } & B(i) \\
{\scriptstyle A(j,i)}\Big\downarrow & {\scriptstyle j \le i} & \Big\downarrow{\scriptstyle B(j,i)} \\
A(j) & \xrightarrow[\ f_j\ ]{} & B(j)
\end{array}
\qquad f_j \circ A(j,i) = B(j,i) \circ f_i
$$

commutes for all comparisons $j \le i$ in S. These arrows are composed in the obvious way, we compose the corresponding functions at each index. Of course, we have to show that the resulting squares commute, and that this composition is associative, but that is straightforward. $\square$

In the previous example we used a poset S to index the constructed category $\widehat{S}$. There is also a more general construction which converts an arbitrary category C into the category $\widehat{C}$ of presheaves of C. We look at that briefly in Section 3.5. Such presheaf categories occur in many places, and are not always recognized as such. For instance, *Set-R* and *Mod-R* are two such categories.

The next example is important in homology.

1.4.2 Example Let R be a fixed ring and consider the category

$$\textbf{\textit{Mod}-R}$$

of right R-modules. If you are not yet happy with *Mod-R* then you can replace it by the category *A Grp* of abelian groups. We construct a new category

$$\textbf{\textit{Ch}}(\textbf{\textit{Mod}-R})$$

out of the objects and arrows of *Mod-R*.

A chain complex, sometimes called a complex, over R is a $\mathbb{Z}$ indexed family

$$A \quad \cdots \longrightarrow A_{n+1} \xrightarrow{\ \alpha_{n+1}\ } A_n \xrightarrow{\ \alpha_n\ } A_{n-1} \longrightarrow \cdots$$

of objects and connecting arrows taken from *Mod-R*. These arrows have to satisfy a certain condition which we come to in a moment.

Note the indexing of the objects. The indexes become smaller as we move along the chain to the right. For this reason it is sometimes convenient to think of the chain as progressing downwards. However, for obvious reasons, a chain is rarely printed in this form.

Also, it is customary to write $d_\bullet$ for each connecting arrow $\alpha_\bullet$ but in the first instance that can be confusing.

The value of each index is important. If we re-index by moving each object 1-step along, then we get a different complex. In particular, the object A_0 plays a special role in the complex. If we change the indexing then that role is give to a different object, and we have a different complex.

The connecting arrows $\alpha_\bullet$ must interact in a simple way.
Given objects B and C of **Mod**-R the zero arrow

$$B \xrightarrow{\quad 0 \quad} C$$

sends each element of B to the zero element of C. (Strictly speaking, we should label each zero arrow with its source and target, but that gets a bit cumbersome.) Consecutive connecting arrows in the complex are required to compose to 0, that is

$$\alpha_n \circ \alpha_{n+1} = 0$$

for each $n \in \mathbb{Z}$.

Each complex A is an object of the new category **Ch**(**Mod**-R). Let's call it an Object to distinguish it from an object of **Mod**-R.

Given two Objects A and B, complexes from **Mod**-R, what is an Arrow

$$A \xrightarrow{\quad f \quad} B$$

in **Ch**(**Mod**-R)? It is an indexed family of arrows of **Mod**-R

$$
\begin{array}{ccc}
A_{n+1} & \xrightarrow{\ f_{n+1}\ } & B_{n+1} \\
\alpha_{n+1}\big\downarrow & & \big\downarrow\beta_{n+1} \\
A_n & \xrightarrow{\ f_n\ } & B_n \\
\alpha_n\big\downarrow & & \big\downarrow\beta_n \\
A_{n-1} & \xrightarrow{\ f_{n-1}\ } & B_{n-1}
\end{array}
$$

such that at each step the corresponding square commutes, that is

$$f_n \circ \alpha_{n+1} = \beta_{n+1} \circ f_{n+1}$$

for each $n \in \mathbb{Z}$. (This is why we choose to write α rather than the customary d for the connecting arrows.)

The structure of $\mathbf{Ch}(\mathbf{Mod}\text{-}R)$ is now more or less obvious.

Given a pair

$$A \xrightarrow{\ f\ } B \xrightarrow{\ g\ } C$$

of Arrows, we have commuting squares

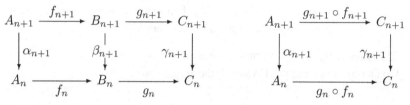

as on the left, and the horizontal components of these compose in $\mathbf{Mod}\text{-}R$ to give commuting squares as on the right. All these composites in $\mathbf{Mod}\text{-}R$ provide the composite

$$A \xrightarrow{\ g \circ f\ } C$$

in $\mathbf{Ch}(\mathbf{Mod}\text{-}R)$.

Verifying that $\mathbf{Ch}(\mathbf{Mod}\text{-}R)$ is a category is now straight forward. It is a simple exercise in diagram chasing which we look at in Section 2.1.   □

The gadget $\mathbf{Ch}(\mathbf{Mod}\text{-}R)$ is central to homology theory. But what is the point of setting it up as a category? It is because some of its properties can be analysed by arrow-theoretic methods without getting inside the internal structure of its Objects and Arrows. This is beyond the scope of this book, but not that far beyond.

Exercises

1.4.1 Try to understand Example 1.4.1. To help with this consider the particular cases where S is a 2-element set partially ordered in the two different ways.

1.4.2 Try to understand Example 1.4.2. To help with this consider the complexes A where only A_{-1}, A_0, A_1 are non-trivial.

1.5 Two simple categories and a bonus

As we are going to see in a moment, every monoid is a category with a simple object structure, and every preset is a category with a simple arrow structure. Every category is a certain kind of amalgam of monoids and presets. Thus whenever you meet a new categorical notion it is worth trying it out on monoids and presets. Sometimes this gives a little bit of insight and sometimes not.

1.5.1 Example Each monoid $(R, \cdot, 1)$ can be viewed as a category with just one object. It doesn't matter what this object is, and it doesn't have any internal structure. Let's use

for this symbolic object. Don't confuse this with the monoid R.

For each $r \in R$ there is an arrow

$$\bigstar \xrightarrow{\quad r \quad} \bigstar$$

and again this has no internal structure. In other words the arrows of the category are the elements of R. Composition of arrows is just the carried operation of R.

The identity arrow

$$id_\bigstar = 1$$

is just the unit of R. This construction does produce a category because the operation on R is associative and 1 is a unit. □

On its own this example is rather trite, but later we will add to it to illustrate several aspects of category theory.

1.5.2 Example Each pre-ordered set $(S, \leq)$ can be viewed as a category. The objects are the elements

$$i, j, k, \ldots$$

of S. Given a pair of objects i, j there is an arrow

$$i \xrightarrow{\qquad} j$$

precisely when $i \leq j$. Thus between any two objects there is at most one arrow. The existence of the arrow indicates a comparison between the objects. It is sometimes convenient to write

$$i \xrightarrow{\;(j,i)\;} j$$

for this arrow. We have

$$id_i = (i,i) \quad \text{since} \quad i \le i$$
$$(k,j) \circ (j,i) = (k,i) \quad \text{since} \quad i \le j \le k \Rightarrow i \le k$$

so the construction does give a category.   □

Again this example looks rather feeble, but again we will add to it later to produce more interesting structures.

In Sections 1.3 and 1.4 we saw various ways of producing a new category out of old categories. There is one very simple example of such a construction. This could have been presented earlier, but we have saved it until the end of this chapter.

1.5.3 Example   Each category C is a collection of objects and a collection of arrows with certain properties. In particular, each arrow

$$A \xrightarrow{\;f\;} B$$

has an assigned source and an assigned target. A formal trick converts C into another category C^{op} called the **opposite** of C. This category C^{op} has the same objects as C. Each arrow f of C, as above, is turned into its formal dual

$$B \xrightarrow{\;f^{\mathrm{op}}\;} A$$

to produce an arrow of C^{op}. The formal composition of these formal arrows is defined by

$$f^{\mathrm{op}} \circ^{\mathrm{op}} g^{\mathrm{op}} = (g \circ f)^{\mathrm{op}}$$

for each composible pair

$$A \xrightarrow{\;f\;} B \xrightarrow{\;g\;} C$$

of arrows from the parent category C. A routine exercise (which you should go through at least once) shows that C^{op} is a category.   □

The process $f \longmapsto f^{\mathrm{op}}$ doesn't actually do anything to the arrows. We merely decide that the words 'source' and 'target' should mean their exact opposites. Thus the change is merely formal rather than actual. This trick shows there is a lot of duality in category theory. Notions often come in dual pairs

dog   god

where a dog of a category C is nothing more than a god of its opposite category C^{op}. We will see many examples of this, and not just when it is raining cats and dogs.

Sometimes the opposite category C^{op} has properties rather different to the parent C. For instance Set^{op} is isomorphic to the category of complete, atomic boolean algebras and complete morphisms. As a simpler version of this the opposite of the category of finite sets is the category of finite boolean algebras. (Both of these observations are instances of Stone duality.)

Exercises

1.5.1 (a) Let R and S be monoids viewed as categories. What is the product category?

(b) Let R and S be presets viewed as categories. What is the product category?

1.5.2 Let S be a preset viewed as a category.

For an arbitrary element $s \in S$ what are the slice categories $(S \downarrow s)$ and $(s \downarrow S)$?

For arbitrary elements $s, t \in S$ what is the butty category $(s \downarrow S \downarrow t)$? Be careful.

1.5.3 Each poset S is a category. What is the opposite S^{op}?

Each monoid R is a category. What is the opposite R^{op}?

1.5.4 Give a short and precise description of the category constructed in Exercise 1.3.3.

2

Basic gadgetry

In this chapter we describe some of the basic gadgets of category theory. We meet notions such as

diagram

monic	epic
split monic	split epic

isomorphism

initial	final

wedge

product	coproduct
equalizer	coequalizer
pullback	pushout

universal solution

some of which are discussed only informally. All of these notions are important, and have to be put somewhere in the book. It is more convenient to have them together in one place, and here seems the 'logical' place to put them. However, that does not mean you should plod through this chapter section by section. I suggest you get a rough idea of the notions involved, and then go to Chapter 3 (which discusses more important ideas). You can come back to this chapter to fill in the missing details.

2.1 Diagram chasing

As in many parts of mathematics, in category theory we sometimes have to show that two things are equal. We don't often, or even ever, have to show that two objects are the same, but we often have to show that two arrows are equal. The main technique for doing that is diagram chasing.

Roughly speaking, a **diagram** in a category is a collection of objects together with a collection of arrows between these objects.

For instance, the following diagram

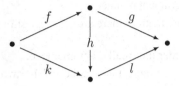

has four objects (unnamed) and five arrows f, g, h, k, l. There are also five composite arrows

$$g \circ f \quad h \circ f \quad l \circ h \circ f \quad l \circ h \quad l \circ k$$

and some of these may be equal. This diagram has three **cells**; the left hand triangle, the right hand triangle, and the outer rectangle (lozenge). Some of these cells may **commute**.

- The left hand triangle commutes if $h \circ f = k$.
- The right hand triangle commutes if $l \circ h = g$.
- The outer cell commutes if $g \circ f = l \circ k$.

Roughly speaking a diagram chase is a process by which we show that a particular cell commutes knowing that other cells commute and using certain other properties of the diagram.

2.1.1 Example   For the diagram above, if the two triangles commute then the outside cell commutes. We are given that

$$h \circ f = k \qquad l \circ h = g$$

and we must show that

$$g \circ f = l \circ k$$

holds. We can do that by equational reasoning. Thus

$$g \circ f = (l \circ h) \circ f = l \circ (h \circ f) = l \circ k$$

is a more or less trivial calculation.

However, it is more common to do this by chasing round the diagram and noting that certain composites are equal. Thus

is what we trace out with our pencil and think whilst we are doing it.   □

Since this example is so trivial it doesn't matter which method we use. For larger diagrams the chase is often easier to explain when talking to someone. This is a bit unfortunate since no-one has yet devised a method of writing down a diagram chase in an efficient manner.

Even with this simple diagram there are other questions to ask.

2.1.2 Example   Consider the small diagram above.

(a) If the outer cell commutes then neither of the two triangles need commute. This is because we could replace h by some other arrow without altering f, g, k, l.

(b) If the outer cell commutes and the right hand triangle commutes, then the left hand triangle need not commute. It is not hard to find an appropriate example in *Set*. Simply let l collapse a lot of elements to the same element.

(c) If the outer cell commutes, the right hand triangle commutes, *and* l is MONIC, then the left hand triangle also commutes. We deal with this in the next section.

(d) If the outer cell commutes and the left hand triangle commutes, then the right hand triangle need not commute. It is not hard to find an appropriate example in *Set*. Simply let the range of f be a small part of its target.

(e) If the outer cell commutes, the left hand triangle commutes, *and* f is EPIC, then the right hand triangle also commutes. We deal with this in the next section.   □

We will take part in many diagram chases. For now I leave you with a couple of simple exercises and one that you might think is a bit devilish.

Exercises

2.1.1   Consider the following diagram.

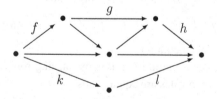

Show that if the four inner triangles commute, then so does the outer cell. Write down the argument in the form of equational reasoning and in a pictorial form of a diagram chase.

2.1.2 Consider the triangular pyramid of arrows.

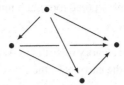

Show that if the three other faces commute then the back face commutes.

2.1.3 Consider a pentagram inscribed in a pentagon.

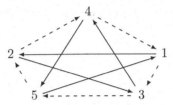

Suppose that the following triangles commute.

$$123 \quad 345 \quad 512 \quad 234 \quad 451$$

Show that a trip twice round the pentagram is equal to a trip once round the pentagon.

2.2 Monics and epics

It is clear, as someone once said, that in the great categorical menagerie all arrows have equal status, but some have more status than others. In this section we look at some of these special arrows.

2.2.1 Definition In a category an arrow

$$B \xrightarrow{\ m\ } A \qquad\qquad A \xrightarrow{\ e\ } B$$

is, respectively,

monic   epic

if for each parallel pair of arrows

$$X \underset{g}{\overset{f}{\rightrightarrows}} B \qquad\qquad B \underset{g}{\overset{f}{\rightrightarrows}} X$$

we have

$$m \circ f = m \circ g \Longrightarrow f = g \qquad f \circ e = g \circ e \Longrightarrow f = g$$

as appropriate. □

What are we getting at here? The following example gives the precursors of monics and epics, but you mustn't read too much into it. Later we will see that it can suggest quite a false story.

2.2.2 Example   Consider a category of structured sets. Each arrow is (carried by) a total function between the carriers of the two objects.

   (m) If

$$B \xrightarrow{\quad m \quad} A$$

is injective as a function then it is monic as an arrow. To see this suppose

$$m \circ f = m \circ g$$

for some parallel pair

$$X \underset{g}{\overset{f}{\rightrightarrows}} B$$

of arrows. We require $f = g$. Thus, since f and g are total functions it suffices to show

$$f(x) = g(x)$$

for each $x \in X$. We have

$$m(f(x)) = (m \circ f)(x) = (m \circ g)(x) = m(g(x))$$

for each such x. But m is injective, that is

$$m(b_1) = m(b_2) \implies b_1 = b_2$$

for $b_1, b_2 \in B$, to give the required result.

   (e) If

$$A \xrightarrow{\quad e \quad} B$$

is surjective as a function then it is epic as an arrow. To see this suppose

$$f \circ e = g \circ e$$

for some parallel pair

$$B \underset{g}{\overset{f}{\rightrightarrows}} X$$

of arrows. We require $f = g$, that is

$$f(b) = g(b)$$

for each $b \in B$. Consider any such $b \in B$. Since e is surjective we have $b = e(a)$ for some $a \in A$. But now

$$f(b) = f(e(a)) = (f \circ e)(a) = (g \circ e)(a) = g(e(a)) = g(b)$$

to give the required result. □

These examples show that

injective $\Longrightarrow$ monic   surjective $\Longrightarrow$ epic

for appropriately nice categories. However, you are warned. Even in nice categories these implications can be far from equivalences. There are several quite common categories of structured sets in which an epic arrow need not be surjective. Roughly speaking an arrow

$$A \xrightarrow{\;e\;} B$$

is epic if the range $e[A]$ of e is a 'large part' of B. Exercises 2.2.5, 2.2.6, and 2.2.8 give examples of this. Of course, in many categories the notions of 'injective arrow' and 'surjective arrow' don't make sense.

Monics and epics are those arrows that can be cancelled on one side or the other. If an arrow has a 1-sided inverse than it can be cancelled on the appropriate side. This gives us special classes of monics and epics.

2.2.3 Definition A pair of arrows

$$B \xrightarrow{\;s\;} A \qquad\qquad A \xrightarrow{\;r\;} B$$

such that

$$r \circ s = id_B$$

are a

section   retraction

respectively (as indicated by the initial letter). □

It is not hard to show that each section is monic and each retraction is epic. For this reason such an arrow is often referred to as a

split monic   split epic

respectively. In some ways this is better terminology.

As we said at the beginning of Section 2.1 rarely do we need to show that two objects of a category are the same. But we often have to show they are isomorphic.

2.2.4 Definition   A pair of arrows

$$B \xrightarrow{\;g\;} A \qquad\qquad A \xrightarrow{\;f\;} B$$

such that

$$g \circ f = id_A \qquad\qquad f \circ g = id_B$$

form an **inverse pair of isomorphisms**. Each arrow is an **isomorphism**.   □

An arrow is an isomorphism if it has a 2-sided inverse, and hence each isomorphism is both a split monic and a split epic. This gives us a short hierarchy of kinds of arrows.

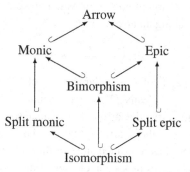

It is easy to see that an arrow that is both a split monic and a split epic is automatically an isomorphism (and there is a stronger result). However, an arrow that is both monic and epic need not be an isomorphism.

2.2.5 Definition   A **bimorphism** is an arrow that is monic and epic.

Each isomorphism is a bimorphism, but there can be bimorphisms which are not isomorphisms.

A category is **balanced** if each bimorphism is an isomorphism.   □

Monics and epics often have a role to play in a diagram chase. There are some exercises to illustrate this.

Exercises

2.2.1   (a) Show that

$$\text{section} \implies \text{monic} \qquad\qquad \text{retraction} \implies \text{epic}$$

$$\text{section} + \text{epic} \implies \text{isomorphism} \qquad \text{retraction} + \text{monic} \implies \text{isomorphism}$$

that is show that if an arrow satisfies the hypothesis then it satisfies the conclu-
sion.

(b) Show that if arrows

$$B \xrightarrow{g} A \xrightarrow{f} B \xrightarrow{h} A$$

satisfy

$$h \circ f = id_A \quad f \circ g = id_B$$

then $g = h$, and each arrow is an isomorphism.

2.2.2 (a) Consider a preset as a category. Show that every arrow is a bimorphism.

(b) When is a poset balanced and when is a preset balanced?

2.2.3 Consider a monoid as a category. Which of the elements (when viewed
as arrows) are monic, epic, a retraction, a section, an isomorphism? When is a
monoid balanced?

2.2.4 Consider a composible pair of arrows.

$$A \xrightarrow{m} B \xrightarrow{n} C$$

Show that if both m and n are monic, then so is the composite $n \circ m$.
Show that if the composite $n \circ m$ is monic, then so is m.
Find an example where the composite $n \circ m$ is monic but n is not.
State the corresponding results for epics.

Obtain similar results (where possible) for the other classes of arrows discussed in this section.

2.2.5 Consider the category *Mon* of monoids, and view $\mathbb{N}$ and $\mathbb{Z}$ as additively written monoids. Show that the insertion

$$\mathbb{N} \xhookrightarrow{e} \mathbb{Z}$$

is epic.

2.2.6 Consider the category *Rng* of rings. Show that the insertion

$$\mathbb{Z} \xhookrightarrow{e} \mathbb{Q}$$

is epic.

2.2.7 (a) Let C be a category of structured sets. Suppose C has a particular object S which has a special element $\star$ (usually not part of the official furnishings) such that for each object A and element $a \in A$, there is a unique arrow

$$S \xrightarrow{\quad \alpha \quad} A$$

with $\alpha(\star) = a$. Show that in C each monic is injective.

(This is a particular instance of a more general notion called a selector, or sometimes a generator.)

(b) Show that in

$$\textbf{\textit{Set, Pos, Top, Mon, Grp, Rng, Set-}}R$$

each monic is injective.

2.2.8 (a) In *Top* an isomorphism is usually called something else. What is the name used?

Show that in *Top* each monic is injective.

Show that an arrow of *Top* that is bijective as a function need not be an isomorphism.

(b) Let *Top*$_2$ be the category of hausdorff spaces and continuous maps.

Show that the insertion

$$\mathbb{Q} \xhookrightarrow{\quad e \quad} \mathbb{R}$$

is epic in this category.

More generally, show that if

$$T \xrightarrow{\quad \epsilon \quad} S$$

is an arrow of *Top*$_2$ where the range $\epsilon[T]$ is dense in the target S, then ϵ is epic.

If you are brave you can show that this result does not hold for *Top*$_1$.

2.2.9 Consider the following cube of arrows $a, b, \ldots, l, m$.

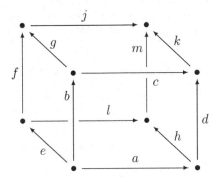

(e) Show that if e is epic and if the other five faces commute, then the back face commutes.

(m) Show that if m is monic and if the other five faces commute, then the bottom face commutes.

2.2.10 In the following diagram suppose the four trapeziums commute.

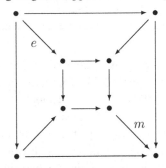

(a) Show that if the inner square commutes then so does the outer square.

(b) Conversely, show that if e is epic, m is monic, and the outer square commutes, then so does the inner square.

2.3 Simple limits and colimits

Limits and colimits (of the categorical kind) occur all over mathematics, and concrete examples of these notions were being used before category theory was invented. Different areas of mathematics tend to use different terminology, mainly for historical reasons rather than natural cussedness, but that is not too important. It was one of the first achievements of category theory to codify and extract the essential content of these notions.

Table 2.1 *Some simple limits and colimits*

	Limit	Template	Colimit
(1)	final object		initial object
(2)	binary product		binary coproduct
(3)	equalizer		co-equalizer
(4)	pullback		
(5)			pushout

In the next four sections we look at the basic examples of these notions, as given in Table 2.1. These examples can be set in a more general context, but we don't attempt that just yet. However, we can look at a particular case of the general notion which you already know about.

Let S be a poset viewed as a category. It is usual to think of the comparison as progressing upwards, that is $i \leq j$ means that i is below j. However, to fit in with the categorical picture we think of the comparison as progressing to the right. Thus

$$i \leq j \qquad\qquad i \longrightarrow j$$

mean the same thing.

Let X be a subset of S. A

left solution   right solution

for X is an element $a \in S$ such that

$$a \leq x \qquad\qquad x \leq a$$

for each $x \in X$. A

limit   colimit

is a 'best possible' solution on the appropriate side. In other words, it is a solution a such that

$$b \leq a \qquad\qquad a \leq b$$

for each solution b of the appropriate handedness.

You should recognize these notions under different names.

Exercises

2.3.1 For a poset what are the notions above usually called?
What happens if X is empty?
What happens if X is a singleton?
What happens if X is a pair of elements?
What differences might occur if S is a preset?

2.4 Initial and final objects

In some categories some objects play special roles because they take up extreme positions.

2.4.1 Definition An object S of a category C is, respectively

initial   final

if for each object A there is a *unique* arrow

$$S \longrightarrow A \qquad\qquad A \longrightarrow S$$

as indicated. Here the uniqueness is important.

Sometimes a final object is said to be terminal.　　　　□

You may not know this terminology, but you already know some examples.

2.4.2 Example (a) Consider the category *Set* of sets and let

$$1 = \{\star\}$$

be a singleton set. For each set A there is a unique arrow

$$A \longrightarrow 1$$

the function that sends everything to $\star$. Thus 1 is a final object of *Set*.

Let $\emptyset$ be the empty set. You will probably have to think about this, but for each set A there is a unique arrow

$$\emptyset \longrightarrow A$$

(since the function doesn't have any requirements that it must satisfy). Thus $\emptyset$ is an initial object of *Set*.

(b) Consider the category *A Grp* of abelian groups. Let O be the trivial group. For each abelian group A, the group O is uniquely embedded in A, and there is a unique morphism

$$A \longrightarrow O$$

to O. Thus O is both initial and final in *A Grp*.   □

A category C may or may not have an initial object. It may or may not have a final object. It can have one without the other. It can have both. If it has both then these objects may or may not be the same. An object that is both initial and final is often called a zero object.

It is easy to show that any two initial objects of a category are uniquely isomorphic. For this reason we usually speak of *the* initial object rather than an initial object. In the same way, any two final objects are uniquely isomorphic, and we speak of *the* final object.

It is common to let 1 be the final object of a category (assuming this exists). Because of certain special cases an arrow

$$1 \overset{a}{\longrightarrow} A$$

to an object A is a global element of A. For instance, in *Set* these pick out the members of a set in the usual sense. In more structured categories these can pick out a special kind of member of an object.

Exercises

2.4.1 Show that in a category any two initial objects are uniquely isomorphic. That is, if I, J are two initial objects, then there is a unique arrow $I \longrightarrow J$, and this is an isomorphism.

State and prove the dual result concerning final objects.

2.4.2 Suppose that I is initial in C. Show that each C-arrow

$$A \longrightarrow I$$

is a retraction. Prove the corresponding result for final objects. Show that if C has an initial object I and a final object F and an arrow

$$F \longrightarrow I$$

then I and F are isomorphic. In such a case we have a zero object.

2.4.3 Show that the category **Pno** has an interesting initial object but a boring final object. What are these objects?

2.4.4 Show that the category **Grp** of groups has both an initial and a final object, and these are the same.

Show that the category **Rng** of unital rings has both an initial and a final object, and these are not the same.

Consider the categories **Idm** and **Fld** of integral domains and fields.

2.4.5 Show that for each set A there is a bijection between elements of A and **Set**-arrows $1 \longrightarrow A$. Show that for each pair of **Set**-arrows

$$A \xrightarrow{\ f\ } B \qquad\qquad 1 \longrightarrow A$$

where the second represents the element $a \in A$, the composite

$$1 \longrightarrow A \xrightarrow{\ f\ } B$$

represents the element $f(a) \in B$.

2.4.6 Let S be a poset and consider the category $\widehat{S}$ of presheaves over S (as described in Example 1.4.1).

(a) Show that this category has a final object $\mathbf{1}$.

(b) Show that for a presheaf $A = (\mathsf{A}, \mathcal{A})$ over S a global element $1 \longrightarrow A$ is a kind of choice function for $\mathcal{A}$. It 'threads' its way through the component sets $\mathsf{A}(s)$. Make precise the notion of 'thread'.

2.5 Products and coproducts

We all know how to form the cartesian product

$$A \times B$$

of two sets A and B, the set of all ordered pairs

$$(a, b)$$

for $a \in A$ and $b \in B$. We also know that often when A and B carry structures of a similar kind, the product $A \times B$ can be furnished with the same kind of structure. Groups, rings, topological spaces, and so on, provide examples of this. In these cases we find that the two projections

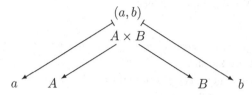

are arrows in the appropriate category.

There is also a dual process which is not so clear.

Given two sets A and B we can form the disjoint union (sum)

$$A + B$$

of the sets, a larger set that includes copies of A and B with minimal interference. Technically, we tag the elements of A and B to remember their origin, and take the union of the tagged versions of the sets.

$$A + B = \left(A \times \{0\}\right) \cup \left(B \times \{1\}\right)$$

We then find that the two embeddings

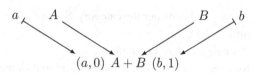

locate disjoint copies of the parent sets within the sum.

What about this dual process for structured sets? Given two groups, or two rings, A and B, can we find a group or ring that includes copies of A and B with minimal interference? It can be done but we have to think a bit before we spot the construction. If you don't know how to do this then you should worry about it for a while.

What we can do here is look at a variant of this dual problem. Given two *abelian* groups A and B we wish to find an *abelian* group that includes copies of A and B with minimal interference. This is easier.

Let's suppose the two abelian groups are written multiplicatively. Thus

$$(A, \cdot, 1) \qquad (B, \cdot, 1)$$

are the two structures. Let

$$A \times B$$

be the cartesian product of these two groups. This, of course, is also an abelian group. We have four morphisms.

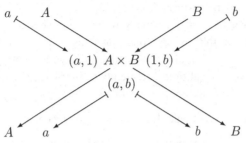

The lower two are the projections. The upper two are the embeddings which solve our problem.

There is something going on here, isn't there? Category theory can help to explain this. In all cases we are looking for a universal solution to a particular kind of problem which comes in two forms, a left handed version and a right handed version.

For the remainder of this section we fix a category C, and we fix a pair of objects A and B of C. We place these as

$$A$$

$$B$$

to help with various diagrams we draw. (Just why we do this will become clear when we look at more general constructions in Chapter 5.)

We are going to look at the left handed version and the right handed version of the problem in parallel. Thus each definition and result that we give is really two definitions or results in one. The left hand side gives the left version and the right hand side gives the right version (in the sense of 'dexterous' not 'correct').

2.5.1 Definition For a pair A, B of objects of a category C, a wedge

to   from

the pair A, B is an object X together with a pair of arrows

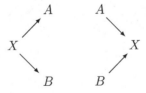

in the parent category C. $\qquad\qquad\qquad\qquad\qquad\qquad\qquad\qquad$ □

Often a wedge of this kind is called a

cone   cocone

Basic gadgetry

depending on which side of the pair it lies. However, it is hardly worth remembering which is which so we call both a wedge.

For a given pair there may be many wedges on one side or the other. We look for a 'best possible' wedge, one that is as 'near' the pair as possible. Technically, we look for a universal wedge. You will probably need to read this next definition several times. Remember also that it is two definitions in one, so in the first instance concentrate on one side.

2.5.2 Definition   Given a pair A, B of objects of a category C, a

product   coproduct

of the pair is a wedge

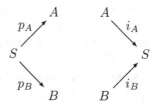

with the following universal property.

For each wedge

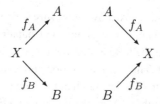

there is a *unique* arrow

$$X \xrightarrow{\ m\ } S \qquad\qquad S \xrightarrow{\ m\ } X$$

such that

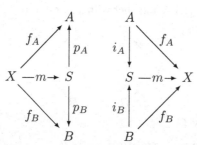

commutes. This arrow m is the **mediating arrow** (or mediator) for the wedge on X.   □

There are a couple of things about this definition that you should notice. Firstly, a product or coproduct is not just an object. It is an object furnished with a pair of arrows. Secondly, the mediator is unique for the given wedge on X. This has some important consequences.

2.5.3 Lemma *Let*

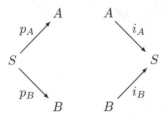

be a

product   coproduct

wedge in the category C. *Let*

$$S \xrightarrow{\quad k \quad} S$$

be any endo-arrow of S for which

$$p_A = p_A \circ k \qquad k \circ i_A = i_A$$
$$p_B = p_B \circ k \qquad k \circ i_B = i_B$$

hold. Then $k = id_S$.

Proof We consider the given wedge both as a special wedge and as an arbitrary wedge. Thus there is a unique arrow, the mediator, such that

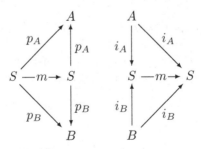

commutes. Since id_S makes these diagram commute we have $m = id_S$. But the arrow k makes this diagram commute, and hence $k = id_S$. □

This result leads to the essential uniqueness of the universal solution.

2.5.4 Lemma   *For objects A, B in a category $\mathbf{C}$ let*

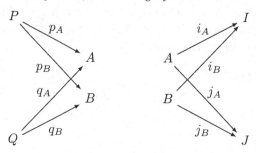

be a pair of

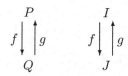

wedges. Then

$$P, Q \qquad\qquad I, J$$

are uniquely isomorphic over the wedges. There are unique arrows

$$
\begin{array}{cc}
P & I \\
f \downarrow \uparrow g & \quad f \downarrow \uparrow g \\
Q & J
\end{array}
$$

such that

(1) $p_A = q_A \circ f \quad p_B = q_B \circ f$ (3) $i_A = g \circ j_A \quad i_B = g \circ j_B$

(2) $q_A = p_A \circ g \quad q_B = p_B \circ g$ (4) $j_A = f \circ i_A \quad j_B = f \circ i_B$

and in particular f and g are an inverse pair of isomorphisms.

Proof   We look at the product, left hand, version and leave the coproduct version as an exercise.

The object Q and the pair q_A, q_B form a product wedge for A, B. The object P and the pair p_A, p_B form an arbitrary wedge for A, B. Thus there is a unique mediator f satisfying (1). By reversing the roles of P and Q there is a unique mediator satisfying (2). From (1) and (2) we have

$$p_A \circ g \circ f = q_A \circ f = p_A$$
$$p_B \circ g \circ f = q_B \circ f = p_B$$

so that a use of Lemma 2.5.3 gives the left hand equality

$$g \circ f = id_P \qquad f \circ g = id_Q$$

and the right hand equality follows by a similar argument.   □

The left hand part of this result shows that if a pair of objects has a product then that gadget is essentially unique. Thus we often speak of *the* product of a pair. Similarly, from the right hand part of this result, we speak of *the* coproduct of a pair of objects.

In some categories not all products or coproducts exist. A pair of objects may have one of these gadgets without the other. The pair may have both, or it may have neither. The existence of products and coproducts in some particular categories is looked at in Exercises 2.5.1 and 2.5.2. Here we look at a result which relates the categorical notions to the concrete construction discussed at the beginning of this section.

2.5.5 Lemma   *Let A and B be a pair of sets. Then the*

cartesian product	disjoint union
$A \times B$	$A + B$

furnished with the canonical functions forms the

| product | coproduct |

of the pair in **Set**.

Proof   We look at the right hand, coproduct, version and leave the left hand version as an exercise.

The elements of

$$A + B$$

are of two kinds

$$(a, 0) \text{ for } a \in A \qquad (b, 1) \text{ for } b \in B$$

where the tag 0 or 1 records the parent of the element. The embeddings

$$
\begin{array}{c}
A \\
i_A \Big\downarrow \qquad i_A(a) = (a, 0) \\
A + B \\
i_B \Big\uparrow \qquad i_B(b) = (b, 1) \\
B
\end{array}
$$

merely tag the input. We must show that these form a coproduct wedge.

Consider any wedge

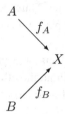

to some set X. We define

$$A + B \xrightarrow{\;m\;} X$$

by

$$m(a,0) = f_A(a) \qquad m(b,1) = f_B(b)$$

for $a \in A$ and $b \in B$. Trivially, the diagram

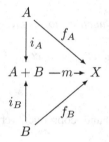

commutes. We must show that m is the only function that makes this diagram commute.

Consider any function

$$A + B \xrightarrow{\;h\;} X$$

with

$$h \circ i_A = f_A \qquad h \circ i_B = f_B$$

that is h makes the diagram commute. For $a \in A$ and $b \in B$ we have

$$h(a,0) = h(i_A(a)) = (h \circ i_A)(a) = f_A(a) = m(a,0)$$
$$h(b,1) = h(i_B(b)) = (h \circ i_B)(b) = f_B(b) = m(b,1)$$

and hence $h = m$, as required. $\qquad\qquad\qquad\qquad\qquad\qquad$ □

We finish this section with a few remarks on terminology and notation. Strictly speaking the two notions we have described here are the

         binary product   binary coproduct

respectively. There are more general notions that deal with an arbitrary number of objects, not just two. The notations

$$A \times B \qquad A + B$$

$$A \, \Pi \, B \qquad A \, \amalg \, B$$

are used for the object associated with the constructed wedge. Of course, a use of '×' does *not* mean that the object is constructed using a cartesian product. In some categories the product and the coproduct produce the same object (but not the same structuring arrows). For such cases

$$A \oplus B$$

is a common notation. This is sometimes called a biproduct.

Exercises

2.5.1 As well as *Top*, choose a selection of the following categories.

Set, *CMon*, *Mon*, *AGrp*, *Grp*, *CRng*, *Rng*, *Set-R*, *Mod-R*, *Pos*

Show that each category has all binary products, and that each is given by a cartesian product with the obvious projections.

2.5.2 Show that each of

Set, *Pos*, *CMon*, *AGrp*, *Set-R*, *Mod-R*, *Top*

has all binary coproducts.
    Can you spot any similarities between the various constructions?

2.5.3 Each poset is a category.
    What is the product of two elements?
    What is the coproduct of two elements?

2.5.4 Show that *Set*$_\perp$ has all binary products and binary coproducts.

2.5.5 Consider the category *SetD* of sets with a distinguished subset.
    Does this category have binary products.
    Does it have binary coproducts.

2.5.6 Consider the category *RelA* of sets and relations of Example 1.3.2. Show this has all binary products and coproducts and give a description of them. (The product is *not* given by a cartesian product.)

2.5.7 Let C be a category with a final object 1 and all binary products.

(a) Show that for each object A the three objects $1 \times A, A, A \times 1$ are isomorphic.

(b) Show that for each triple A, B, C of objects, the two objects

$$(A \times B) \times C \quad A \times (B \times C)$$

are isomorphic.

2.5.8 Let C be a category with all binary products and coproducts. For objects A, B, C let

$$L = A \times C + B \times C \quad R = (A + B) \times C$$

to form two more objects. Show there is an arrow

$$L \longrightarrow R$$

and find an example to show that there need not be an arrow $R \longrightarrow L$.

2.5.9 In the category *AGrp* of abelian groups the cartesian product of two objects implements both the product and the coproduct. Does this work in *Grp*?

Consider the cartesian product $A \times B$ of two abelian groups. This gives the coproduct of A and B in *AGrp*. Does this give the coproduct of A and B in *Grp*?

2.6 Equalizers and coequalizers

When we first see their categorical definition, equalizers and coequalizers are not something we immediately relate to our previous mathematical experience. They are a couple of notions which help in certain categorical situations. However, once we become familiar with the idea we begin to realize that we have seen particular instances of the notions.

There are two notions here, the left notion and the right notion. We develop the two versions in parallel. For instance, the following definition is two definitions in one. As with other parallel developments, at a first reading concentrate on one side. Once you understand that come back and do the other side. In this instance I suggest that the left hand, equalizing, side is easier to begin with.

2.6.1 Definition Given a parallel pair

$$A \underset{g}{\overset{f}{\rightrightarrows}} B$$

of arrows in a category C, an arrow

$$X \xrightarrow{\quad h \quad} A \qquad\qquad B \xrightarrow{\quad h \quad} X$$

makes equal the parallel pair if

$$f \circ h = g \circ h \qquad\qquad h \circ f = h \circ g$$

that is, the two composite arrows

$$X \xrightarrow[\ g \circ h\ ]{\ f \circ h\ } B \qquad\qquad A \xrightarrow[\ h \circ g\ ]{\ h \circ f\ } X$$

agree. □

Any given parallel pair could be made equal, on one side or the other, by many different arrows. We look for a 'best possible' coalescing arrow.

2.6.2 Definition   Given a parallel pair

$$A \xrightarrow[g]{\ f\ } B$$

of arrows in a category C,

| an equalizer | a coequalizer |

is an arrow

$$S \xrightarrow{\quad k \quad} A \qquad\qquad B \xrightarrow{\quad k \quad} S$$

which makes equal f and g, and has the following universal property.
  For each arrow

$$X \xrightarrow{\quad h \quad} A \qquad\qquad B \xrightarrow{\quad h \quad} X$$

which makes equal the parallel pair, there is a *unique* arrow

$$X \xrightarrow{\quad m \quad} S \qquad\qquad S \xrightarrow{\quad m \quad} X$$

such that

$$h = k \circ m \qquad\qquad h = m \circ k$$

holds. This m is the **mediating arrow** (or mediator) for the arrow h. □

Read this definition a couple of times and compare it with Definition 2.5.2. Later, in Chapter 4, we will see that both notions are particular instances of a more general notion.

For now we develop the idea of Definition 2.6.2. We follow a path quite similar to that in Section 2.5. Here is the analogue of Lemma 2.5.3.

2.6.3 Lemma   *Each equalizer is monic. Each coequalizer is epic.*

The proof of this is similar to that of Lemma 2.5.3, so we leave it as an exercise. We use the result to obtain the analogue of Lemma 2.5.4.

2.6.4 Lemma   *For a parallel pair*

$$A \underset{g}{\overset{f}{\rightrightarrows}} B$$

of arrows in a category C *let*

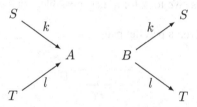

be a pair of

equalizers   coequalizers

respectively. Then S, T *are uniquely isomorphic over the wedges, In other words, there are unique arrows*

such that

(1) $\quad l = k \circ m$ \qquad (3) $\quad l = m \circ k$

(2) $\quad k = l \circ n$ \qquad (4) $\quad k = n \circ l$

and in particular m *and* n *are an inverse pair of isomorphisms.*

Proof   We look at the coequalizer, right hand, version and leave the equalizer version as an exercise.

The arrow l makes equal f and g. The arrow k is the coequalizer of f and g. Thus there is a unique mediator m satisfying (3). By reversing the roles of l

and k we see there is a unique mediator n satisfying (4). From (3) and (4) we have

$$n \circ m \circ k = n \circ l = k = id_S \circ k$$

and hence

$$n \circ m = id_S$$

since k is epic. Similarly

$$m \circ n = id_T$$

to show that m and n are an inverse pair of isomorphisms. $\square$

The left hand part of this result shows that if a pair of arrows has an equalizer then that gadget is essentially unique. Thus we speak of *the* equalizer of a pair. Similarly, from the right hand part of this result, we speak of *the* coequalizer of a pair of arrows.

Let's now look at a few examples. Any given pair of arrows need not have an equalizer, nor a coequalizer. In contrast to this for some categories these gadgets always exist.

2.6.5 Example Let

$$A \overset{f}{\underset{g}{\rightrightarrows}} B$$

be a parallel pair of functions, arrows in *Set*. Let

$$S = \{a \in A \mid f(a) = g(a)\}$$

be the set of elements of A on which f and g agree. Then the insertion

$$S \overset{i}{\hookrightarrow} A$$

is the equalizer of f and g. To see that suppose the function

$$X \overset{h}{\longrightarrow} A$$

makes equal f and g. For each $x \in X$ we have

$$f(h(x)) = (f \circ h)(x) = (g \circ h)(x) = g(h(x))$$

so that $h(x) \in S$, and hence the function

$$X \overset{m}{\longrightarrow} S$$
$$x \longmapsto h(x)$$

is the required mediator. $\square$

A similar idea can be used in several other categories.

The category *Set* also has all coequalizers. To obtain these we combine two standard constructions which, at first sight, seem to have little to do with category theory. Almost certainly you will know the content of the following example, but you may not have seen it set out like this.

2.6.6 Example   Let S be an arbitrary set, and let $\sim$ be an equivalence relation on S. This relation partitions S into blocks (equivalence classes). For each $s \in S$ let $[s]$ be the block in which s lives, and let

$$S/\sim$$

be the set of all such blocks. Let

$$S \xrightarrow{\;\sigma\;} S/\sim$$
$$s \longmapsto [s]$$

be the induced surjection.

Let

$$S \xrightarrow{\;h\;} X$$

be any function. The **kernel** of h is the relation $\approx$ on S given by

$$s_2 \approx s_2 \iff h(s_1) = h(s_2)$$

for $s_1, s_2 \in S$. Trivially, this is an equivalence relation.

Now suppose $\approx$ includes $\sim$, that is

$$s_1 \sim s_2 \implies h(s_1) = h(s_2)$$

for $s_1, s_2 \in S$. Under these conditions there is a commuting triangle

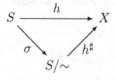

for some unique function $h^\sharp$. This function is given by

$$h^\sharp([s]) = h(s)$$

for $s \in S$. The only problem is to show that $h^\sharp$ is well-defined. $\square$

To produce a coequalizer we generate a certain equivalence relation.

2.6.7 Example Let

$$A \underset{g}{\overset{f}{\rightrightarrows}} B$$

be a parallel pair of functions, arrows in Set. Let $\rightsquigarrow$ be the relation on B given by

$$b_1 \rightsquigarrow b_2 \iff (\exists a \in A)[b_1 = f(a) \text{ and } b_2 = g(a)]$$

for $b_1, b_2 \in B$. Let $\sim$ be the equivalence relation on B generated by $\rightsquigarrow$. (An explicit description of $\sim$ is not as easy as it looks.)

Using the construction of Example 2.6.6 we may check that the canonical quotient

$$B \longrightarrow B/\sim$$

is the coequalizer of the pair f, g (in Set). □

A variation of this construction can be used in some Set-based categories. We first pass down to Set, produce a quotient set, and then furnish this to produce an object of the parent category.

Exercises

2.6.1 Prove Lemma 2.6.3, and complete the proof of Lemma 2.6.4.

2.6.2 Complete the proof of Example 2.6.5. In other words, show that the function m does make the relevant triangle commute, and it is the only function to make that triangle commute.

2.6.3 Consider a parallel pair of morphisms

$$A \underset{g}{\overset{f}{\rightrightarrows}} B$$

between groups (written multiplicatively).

(a) Let

$$E = \{a \in A \mid f(a) = g(a)\}$$

and let S be the subgroup of A generated by E. Show that the insertion

$$S \lhook\joinrel\longrightarrow A$$

is the equalizer of f and g in Grp.

62 Basic gadgetry

(b) Let

$$F = \{f(a)g(a)^{-1} \,|\, a \in A\}$$

and let K be the normal subgroup generated by F. Show that the canonical quotient

$$B \longrightarrow B/K$$

is the coequalizer of f and g in **Grp**.

2.6.4 Write down the details missing from Example 2.6.6. (None of these details are difficult, but you should at least list what is missing.)

2.6.5 Write down the details missing from Example 2.6.7.

2.6.6 Let

$$S \overset{\phi}{\underset{\psi}{\rightrightarrows}} T$$

be a parallel pair of continuous maps between topological spaces. Let

$$T \overset{\theta}{\longrightarrow} T/{\sim}$$

be the coequalizer in **Set** of the pair of functions ϕ and ψ.

Show there is a suitable topology on $T/{\sim}$ for which θ becomes the coequalizer of the pair ϕ, ψ in **Top**.

2.6.7 Consider the forgetful functor

$$\textbf{Pre} \longleftarrow \textbf{Pos}$$

from posets to presets. Eventually we will see that this exercise produces the left adjoint to this functor.

(a) Let S be a pre-ordered set and let $\sim$ be the relation on S given by

$$a \sim b \iff a \leq b \leq a$$

(for $a, b \in S$).

Show that $\sim$ is an equivalence relation on the set S.

Show that S is a poset precisely when $\sim$ is equality.

(b) Let $S/{\sim}$ be the set of blocks of $\sim$ and let

$$S \overset{\eta}{\longrightarrow} S/{\sim}$$

be the canonical quotient.

Show that letting

$$[a] \le [b] \iff a \le b$$

for $a, b \in S$ produces a well-defined partial order on $S/\sim$.

Show that the function η is monotone.

(c) Consider any monotone map

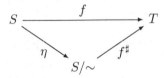

from the preset S to a poset T.

Show that

$$a \sim b \implies f(a) = f(b)$$

for all $a, b \in S$.

Show there is a unique monotone map $f^\sharp$ such that the

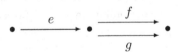

triangle commutes.

2.6.8 Consider a diagram

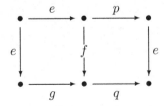

where e makes equal f and g. Suppose also there is a commuting diagram

where the bottom and top composites are identity arrows.

Show that e is the equalizer of f and g.

2.7 Pullbacks and pushouts

As I said in Section 2.3, the notions discussed in Sections 2.4, 2.5, 2.6, and this section are particular cases of a more general notion, that of a

limit   colimit

of a diagram. In this section we begin to use the terminology and the ideas behind this more general notion. This is not essential here, but it will help when we look at the more general notion in Chapter 5.

Each of the gadgets we are interested in is the universal solution of a problem posed by a diagram. For the simple gadgets of this chapter the shape of the diagram - the template - determines the name of the gadget. These templates are given in Table 2.1 on page 44 with the names

left universal solution   right universal solution

for that shape. (The template for row (1) is there, but it's empty.)
    The diagram for a

pullback   pushout

is a

left wedge   right wedge

as in the table. This wedge poses a problem on the appropriate side.
    As before, we develop the two notions in parallel. So each definition or result is two for the price of one.

2.7.1 Definition   Let

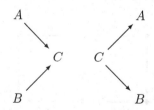

be a wedge in a category C. A solution for the

left   right

problem posed by the wedge is a wedge

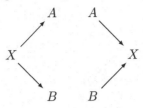

(of the opposite handedness) such that the square

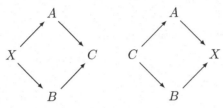

commutes. ☐

Notice that we haven't given each arrow a name. We are beginning to work more and more in terms of diagrams, and we name an arrow only when it becomes necessary. (It is also the case that we need not name the objects, but let's not go that far just yet.)

The problem posed by a wedge can have many different solutions. We look for a 'best possible' solution.

2.7.2 Definition Let

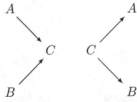

be a wedge in a category C. A

pullback	pushout

for the wedge is a solution

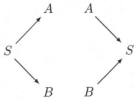

with the following universal property.

For each solution

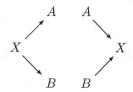

there is a *unique* arrow

such that the following diagram commutes.

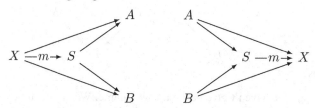

This arrow m is the **mediating arrow** (or mediator) for the wedge on X.   □

As always, each universal solution is essentially unique. To prove this here we first obtain the analogue of Lemma 2.5.3.

2.7.3 Lemma   *Let*

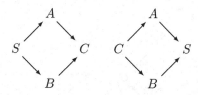

be a

           pullback   *pushout*

square in the category ***C***. *Let*

$$S \xrightarrow{\ k\ } S$$

be any endo-arrow of S for which the two triangles

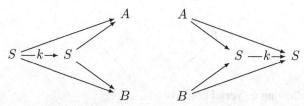

commute. Then $k = id_S$.

You should be able to see the proof of this immediately. We use the result to obtain the analogue of Lemma 2.5.4.

2.7.4 Lemma   *For a wedge*

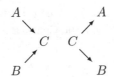

in a category **C**, *let*

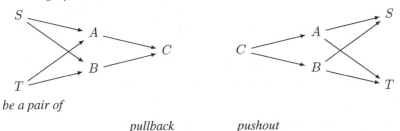

be a pair of

pullback   pushout

squares. Then S, T are uniquely isomorphic over the parent wedge. In other words, there are unique arrows

such that all the triangles

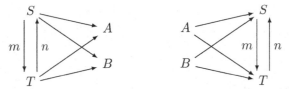

commute. In particular, m and n are an inverse pair of isomorphisms.

Proof   We are given two solutions of the parent problem. Furthermore, each is a universal solution. Thus the associated mediators are the arrows m and n. We now apply Lemma 2.7.3 to the two compounds

$$n \circ m \qquad m \circ n$$

to show these are

$$id_S \qquad id_T$$

respectively.   □

If you found this proof a little hard to follow, try labelling the arrows and re-work the argument using equational reasoning.

Let's now look at a couple of examples of these notions.

2.7.5 Examples   (a) In the category *Set* (of sets and functions) consider a wedge of functions f, g as on the right. Consider also the product

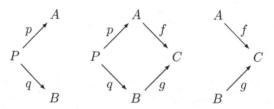

wedge p, q of the two two sets A, B, as on the left. This gives us a square of arrows, as in the centre, but this square need *not* commute.

Let

$$S = \{z \in P \mid f(p(z)) = g(q(z))\}$$

be the set of elements of P which arrive at the same place no matter which route they take. Let

$$S \overset{i}{\hookrightarrow} P$$

be the insertion of S in P. Then the wedge

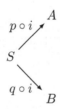

is the pullback of the parent wedge.

To see this observe first that, by construction, this wedge on S is a solution to the problem posed by the parent wedge.

Consider any solution to the posed problem.

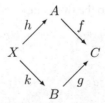

Using the product property we have a commuting diagram

$$
\begin{array}{ccc}
 & & A \\
 & {\scriptstyle h}\nearrow & \uparrow {\scriptstyle p} \\
X & \xrightarrow{\ l\ } & P \\
 & {\scriptstyle k}\searrow & \downarrow {\scriptstyle q} \\
 & & B
\end{array}
$$

for some unique function l. For each $x \in X$ we have

$$
\begin{aligned}
f(p(l(x))) &= (f \circ p \circ l)(x) \\
&= (f \circ h)(x) \\
&= (g \circ k)(x) \\
&= (g \circ q \circ l)(x) = g(q(l(x)))
\end{aligned}
$$

to show that $l(x) \in S$. We may now check that

$$
\begin{array}{ccc}
X & \xrightarrow{\ \ l\ \ } & S \\
x & \longmapsto & l(x)
\end{array}
$$

is the required unique mediating arrow.

(b) In the category *Set* (of sets and functions) consider a wedge of functions f, g as on the left. Consider also the coproduct wedge i, j of

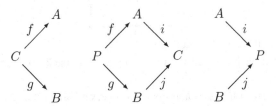

the two two sets A, B, as on the right. This gives us a square of arrows, as in the centre, but this square need *not* commute.

Let $\rightsquigarrow$ be the relation on P given by

$$
z_1 \rightsquigarrow z_2 \iff (\exists c \in C)[z_1 = (i \circ f)(c) \text{ and } z_2 = (j \circ g)(c)]
$$

for $z_1, z_2 \in B$. Let $\sim$ be the equivalence relation on P generated by $\rightsquigarrow$. Let $S = P/\sim$ and let

$$
P \xrightarrow{\ \ k\ \ } S
$$

be the canonical quotient. By construction the square

commutes, and so we do have a solution to the posed problem. We need to show it is a universal solution. This follows by a few calculations. □

Did you spot anything about these two constructions?

Exercises

2.7.1 (a) Suppose the category C has all binary products and all equalizers. Show that C has all pullbacks.

(b) Suppose the category C has all binary coproducts and all coequalizers. Show that C has all pushouts.

2.7.2 Let S be a poset which as a category has all pushouts. What does this mean lattice theoretically. (There is a lattice theoretic notion which matches the categorical notion, but is rarely recognized as such.)

2.7.3 Consider the following commuting diagram

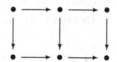

of two inner cells and one outer cell.

Show that if each of the two inner cells is a pullback, then so is the outer cell.

Show that if the outer cell and the right inner cell are pullbacks, then the left inner cell is a pullback.

Sort out the corresponding results for pushouts.

2.7.4 Show that monics are stable across pullbacks, that is if

is a pullback and f is monic, then h is monic.

2.7.5 Show that equalizers are stable across pullbacks, that is if

is a pullback and f is is the equalizer of some pair, then h is the equalizer of some other pair.

2.8 Using the opposite category

In Sections 2.2 to 2.7 we have looked at six pairs of gadgets, a left version and a right version. By using the opposite category C^{op} we can make precise this left-right symmetry, and halve the work.

Consider any arrow

of a category C. Then

$$f \text{ is monic in } C \Longleftrightarrow f^{\text{op}} \text{ is epic in } C^{\text{op}}$$
$$f \text{ is epic in } C \Longleftrightarrow f^{\text{op}} \text{ is monic in } C^{\text{op}}$$

to show that one of the notions immediately gives the other one.

Consider any object

$$K$$

of a category C. Then

$$K \text{ is final in } C \Longleftrightarrow K \text{ is initial in } C^{\text{op}}$$
$$K \text{ is initial in } C \Longleftrightarrow K \text{ is final in } C^{\text{op}}$$

to show that one of the notions immediately gives the other one.

This duality is a useful trick. It can help to save work. For instance, we have seen that each gadget discussed in this chapter is 'essentially unique', but in each case we only did half the proof. This is because the other half is the same argument carried out in the opposite category.

Exercises

2.8.1 Check that each of the gadgets of this chapter is the dual of a similar gadget in the opposite category.

3

Functors and natural transformations

Eilenberg and MacLane invented (discovered) category theory in the early 1940s. They were working on Čech cohomology and wanted to separate the routine manipulations from those with more specific content. It turned out that category theory is good at that. Hence its other name abstract nonsense which is not always used with affection.

Another part of their motivation was to try to explain why certain 'natural' constructions are natural, and other constructions are not. Such 'natural' constructions are now called natural transformations, a term that was used informally at the time but now has a precise definition. They observed that a natural transformation passes between two gadgets. These had to be made precise, and are now called functors. In turn each functor passes between two gadgets, which are now called categories. In other words, categories were invented to support functors, and these were invented to support natural transformations.

But why the somewhat curious terminology? This is explained on pages 29 and 30 of Mac Lane (1998).

... the discovery of ideas as general as these is chiefly the willingness to make a brash or speculative abstraction, in this case supported by the pleasure of purloining words from philosophers: "Category" from Aristotle and Kant, "Functor" from Carnap ...

That, of course, is the bowdlerized version.

Most of the basic notions were set up in Eilenberg and MacLane (1945) and that paper is still worth reading.

In this chapter we look at these two basic notions. We deal first with the definition of functor, and then look at various examples of these gadgets. After that we look at the definition of natural transformation and conclude with several examples of these gadgets.

3.1 Functors defined

The basic belief of category theory is that whenever we conceive of a collection of 'objects' - things we don't want to take apart - we should, at the same time, decide how these 'objects' are to be compared. We then formalize a category. In other words, for any given category C we should think of the arrows of C as those gadgets which compare the objects. Furthermore, these arrows are just as important as, and sometimes more important than, the objects. To stay true to this principle we must now ask a question. We have invented a collection of things called categories. How should categories be compared? Functors are the comparison gadgets.

3.1.1 Definition   (Preliminary) Given a pair of categories

$$Src \qquad Trg$$

a functor

$$
\begin{array}{ccc}
Src & \longrightarrow & Trg \\
A & \longmapsto & FA \\
f & \longmapsto & F(f)
\end{array}
$$

consists of two assignments. One sends objects to objects, and the other sends arrows to arrows. □

As here, it is customary to use the same letter for both assignments. I find it helpful to use brackets in the arrow assignment but not in the object assignment.

Of course, there is more to a functor than just a pair of assignments. It is supposed to be a 'morphism of categories' in the sense that it must respect the structure of the two categories. What can that mean?

The first bit is that a functor F must preserve identity arrows. For each Src object A we must have

$$ A \xrightarrow{\ id_A\ } A \quad \longmapsto \quad FA \xrightarrow{\ id_{FA}\ } FA $$

that is

$$F(id_A) = id_{FA}$$

in equational form.

That part is easy, but now comes the part that might be confusing.

The second bit is that a functor F must preserve composition of composible arrows. But here there can be a twist in the tale. Given an arrow

$$A \xrightarrow{\ f\ } B$$

in the source category **Src**, the arrow $F(f)$ in the target category **Trg** must pass between the two objects FA and FB of **Trg**. But there are two ways it might do that. It can preserve the direction or it can reverse the direction. This leads to two kinds of functors.

Covariant

$$A \xrightarrow{\ f\ } B \quad\longmapsto\quad FA \xrightarrow{\ F(f)\ } FB$$

Contravariant

$$A \xrightarrow{\ f\ } B \quad\longmapsto\quad FB \xrightarrow{\ F(f)\ } FA$$

For both kinds the source and target of an arrow are preserved as an unordered pair. For a covariant functor the direction of the arrow is always preserved, but for a contravariant functor the direction of the arrow is always reversed.

Notice that the direction is *not* sometimes preserved and sometimes reversed. It is always one or the other.

3.1.2 Definition   (In full) Given a pair of categories

$$\textbf{Src} \qquad \textbf{Trg}$$

a functor

$$\begin{array}{l} \textbf{Src} \longrightarrow \textbf{Trg} \\ A \longmapsto FA \\ f \longmapsto F(f) \end{array}$$

consists of two assignments. One sends objects to objects, and the other sends arrows to arrows.

(Co) For a **covariant** functor composition is preserved as follows.

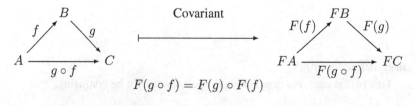

$$F(g \circ f) = F(g) \circ F(f)$$

(Contra) For a **contravariant** functor composition is preserved as follows.

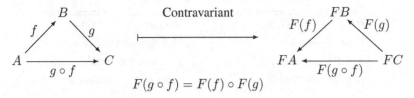

$$F(g \circ f) = F(f) \circ F(g)$$

For both kinds identity arrows are preserved in the sense that

$$F(id_A) = id_{FA}$$

for each object A. □

As mentioned before this definition, it doesn't make sense to say a functor is sometimes covariant and sometimes contravariant.

(There is a notion of a multi-functor with several input positions for objects. Such multi-functors can be covariant in some input positions and contravariant in the other input positions. The simplest example of this is the 2-placed hom-functor. We meet this in the next section.)

In the main we deal with covariant functors and refer to these as functors. Only when it is important do we specifically mention the variance of a functor.

Exercises

3.1.1 Consider a pair S, T of monoids viewed as categories.
What is a covariant functor from S to T?
What is a contravariant functor from S to T?

3.1.2 Consider a pair S, T of presets viewed as categories.
What is a covariant functor from S to T?
What is a contravariant functor from S to T?

3.1.3 Show that for each pair *Src* and *Trg* of categories, covariant functors

$$Src^{\mathrm{op}} \longrightarrow Trg \qquad Src \longrightarrow Trg^{\mathrm{op}}$$

are just contravariant functors from *Src* to *Trg*.

3.1.4 Define the composite $G \circ F$ of two functors F and G (perhaps of different variance), and show that the result is a functor.
How does the variance of $G \circ F$ relate to that of F and G?

Table 3.1 *Some forgetful functors*

$Rng \longrightarrow AGrp$	forget the $\times$-structure
$Rng \longrightarrow Mon$	forget the +-structure
$Mod\text{-}R \longrightarrow AGrp$	forget the action
$Mod\text{-}R \longrightarrow Set\text{-}R$	forget the +-structure
$CMon \longrightarrow Mon$	For all three
$AGrp \longrightarrow Grp$	the commutative property
$CRng \longrightarrow Rng$	is forgotten
$Sup \longrightarrow Join \longrightarrow Pos$	first forget arbitrary suprema but retain joins then forget these
$Inf \longrightarrow Meet \longrightarrow Pos$	

3.2 Some simple functors

In this section we look at some simple examples of functors. Most of these are chosen merely to illustrate the notion, but one or two are important in their own right.

Forgetful functors

Let C be any category of structured sets. Thus each object

$$(A, \cdots)$$

is a set furnished with some gadgetry, and each arrow

$$(A, \cdots) \longrightarrow (B, \cdots)$$

is a function between the two carrying sets. Arrow composition is just function composition. Here we have a covariant functor

$$C \longrightarrow Set$$

which sends each object to its carrying set, and each arrow to its carrying function. I know this is not very exciting, but the idea can help to clear up a bit of confusion from time to time.

This is an example of a forgetful functor. There are a few more given in Table 3.1. For each of these something is forgotten (or ignored) as we pass from the source category to the target category. In the first batch some structure is forgotten. In the second batch some property is forgotten. In the third batch it is a mixture of structure and property that is forgotten.

All of these forgetful functors are covariant. Occasionally we meet a contravariant forgetful functor. Consider the functors

$$Pos^{\dashv} \xrightarrow{\;\;L\;\;} Pos \qquad\qquad Pos^{\dashv} \xrightarrow{\;\;R\;\;} Pos$$

which pick out the left component and the right component of each arrow.

Hom-functors

The next two examples, one covariant and one contravariant, are very important. We will meet them many times in several forms.

Let C be an arbitrary category. Let K be an object of C. For each object A we have an arrow set

$$LA = C[K, A] \qquad RA = C[A, K]$$

(unless, of course, this collection is too big to be a set). Thus we have a pair of object assignments

$$
\begin{array}{ccc}
C & \longrightarrow & Set \\
A & \longmapsto & LA
\end{array}
\qquad
\begin{array}{ccc}
C & \longrightarrow & Set \\
A & \longmapsto & RA
\end{array}
$$

to the category of sets. These are the object assignments of a pair of functors where each arrow

$$A \xrightarrow{\;\;f\;\;} B$$

of C is sent to

$$
\begin{array}{ccc}
C[K, A] & \xrightarrow{\;\;L(f)\;\;} & C[K, B] \\
r & \longmapsto & f \circ r
\end{array}
\qquad
\begin{array}{ccc}
C[B, K] & \xrightarrow{\;\;R(f)\;\;} & C[A, K] \\
l & \longmapsto & l \circ f
\end{array}
$$

that is

$$L(f)(r) = f \circ r \qquad\qquad R(f)(l) = l \circ f$$

respectively. Each of these is a

covariant   contravariant

hom-functor, respectively.

Of course, we should check that we do have a pair of functors, and sort out the variance of each. This is not difficult but it is worth setting down the two calculations in parallel.

Consider a pair of arrows

$$A \xrightarrow{\ f\ } B \xrightarrow{\ g\ } C$$

of C. Consider also arrows

$$K \xrightarrow{\ r\ } A \qquad\qquad C \xrightarrow{\ l\ } K$$
$$r \in LA \qquad\qquad\qquad l \in RC$$

respectively. Then

$$
\begin{aligned}
\bigl(L(g) \circ L(f)\bigr)(r) &= L(g)\bigl(L(f)(r)\bigr) \\
&= L(g)(f \circ r) \\
&= g \circ (f \circ r) \\
&= (g \circ f) \circ r \\
&= L(g \circ f)(r)
\end{aligned}
\qquad
\begin{aligned}
\bigl(R(f) \circ R(g)\bigr)(l) &= R(f)\bigl(R(g)(l)\bigr) \\
&= R(f)(l \circ g) \\
&= (l \circ g) \circ f \\
&= l \circ (g \circ f) \\
&= R(g \circ f)(l)
\end{aligned}
$$

to show that L is a covariant functor and R is a contravariant functor. We should also show that

$$L(id_A) = id_{LA} \qquad R(id_A) = id_{RA}$$

but that is more or less trivial.

Exercises

3.2.1 For an arbitrary category C consider the arrow category $C^{\downarrow}$ of Example 1.3.5. Show there are three functors

$$C^{\downarrow} \underset{T}{\overset{S}{\rightrightarrows}} C \xrightarrow{\ \Delta\ } C^{\downarrow}$$

between the categories. The functor Δ is called the **diagonal** functor.

3.2.2 Let S be a poset viewed as a category. What is a contravariant functor

$$S \longrightarrow \mathbf{Set}$$

to $\mathbf{Set}$? You have seen this notion before.

3.2.3 Let R be a monoid viewed as a category. What is a covariant functor, and what is a contravariant functor

$$R \longrightarrow \mathbf{Set}$$

to $\mathbf{Set}$? Both these notions occur elsewhere in this book, but are described in a different way.

3.2.4 In Example 1.3.2 we looked at the graph $\Gamma(f)$ of a function (between sets). Show that this is the arrow assignment of a functor, and determine the variance of that functor.

3.2.5 In Example 1.3.4 we saw how to produce the product $C \times D$ of two categories. This enables us to think of 2-placed functors

$$C \times D \longrightarrow Trg$$

with two inputs. In particular, for a given category C we can view

$$C \times C \xrightarrow{\;C[-,-]\;} Set$$

as a 2-placed functor. Think about this, and what it should mean.

3.3 Some less simple functors

In this section we look at some examples of functors with a bit more content, although none of them are very complicated. Some of the examples may look a bit contrived, but each one is a miniature version of something quite important.

3.3.1 Three power set functors

It may come as a surprise, but different functors can have the same object assignment. In this block we look at three endo-functors on Set

$$Set \longrightarrow Set$$

where the object assignment of each sends a set A to its power set.

$$A \longmapsto \mathcal{P}A$$

Furthermore, two of these functors are covariant and one contravariant.

It is common to use the same letter as the name of both the object assignment and the arrow assignment. Here we can't do that. We use

$$Set \underset{\forall}{\overset{\exists}{\rightleftarrows\!\!\!-\!\!\!\rightarrows}} Set$$

as the three names, where the two outer ones are covariant and the central one is contravariant. The stacking of the functors is significant, but that won't become clear for some time. Also the use of '$\exists$' and '$\forall$' may look a bit pretentious, but in a more general setting these functors really do have something to do with quantification. We will see just a hint of this shortly.

For each set A we have

$$\exists A = \mathcal{P}A \qquad \mathsf{I}A = \mathcal{P}A \qquad \forall A = \mathcal{P}A$$

as the three object assignments.

For the three arrow assignments consider any arrow of *Set*

$$A \xrightarrow{\ f\ } B$$

a function between the two sets. We require three functions

$$\mathcal{P}A \xrightarrow{\ \exists(f)\ } \mathcal{P}B \qquad \mathcal{P}A \xleftarrow{\ \mathsf{I}(f)\ } \mathcal{P}B \qquad \mathcal{P}A \xrightarrow{\ \forall(f)\ } \mathcal{P}B$$

where the central one reverses the direction. We set

$$\exists(f)(X) = f[X] \qquad \mathsf{I}(f)(Y) = f^{\leftarrow}(Y) \qquad \forall(f)(X) = f[X']'$$

for each $X \in \mathcal{P}A$ and $Y \in \mathcal{P}B$. Here $f[\cdot]$ gives the direct image across f, and $f^{\leftarrow}(\cdot)$ gives the inverse image across f. Notice that $\forall(f)$ uses the dual complement of the direct image (for $(\cdot)'$ is complementation).

We find that

$$b \in \exists(f)(X) \iff (\exists a \in A)[b = f(a) \ \& \ a \in X]$$
$$b \in \forall(f)(X) \iff (\forall a \in A)[b = f(a) \Rightarrow a \in X]$$

for all $X \in \mathcal{P}$ and $b \in B$. Notice how the description matches the name. We also have

$$a \in \mathsf{I}(f)(Y) \iff f(a) \in Y$$

for all $Y \in \mathcal{P}B$ and $a \in A$.

It is not immediately clear that these constructions do give functors, so we must check that.

For functions

$$A \xrightarrow{\ f\ } B \xrightarrow{\ g\ } C$$

we must show that

$$\exists(g \circ f) = \exists(g) \circ \exists(f) \qquad \mathsf{I}(g \circ f) = \mathsf{I}(f) \circ \mathsf{I}(g) \qquad \forall(g \circ f) = \forall(g) \circ \forall(f)$$

and the identity requirements. This is not hard, if you take a bit of care.

Exercises

3.3.1 Consider the three constructions $\exists, !, \forall$ on *Set*. Show that each passes across composition in the required manner.

3.3.2 For each set A the power set $\mathcal{P}A$ is a poset under inclusion. Show that for each function f, as on the left,

$$A \xrightarrow{\quad f \quad} B \qquad \mathcal{P}A \xleftarrow{\quad !(f) \quad} \mathcal{P}B$$

with $\exists(f)$ above and $\forall(f)$ below the $!(f)$ arrow

the three functions on the right form a double poset adjunction.

3.3.2 Spaces, presets, and posets

In this block we compare the category *Top* of topological spaces with the categories *Pre* and *Pos*. We set up four functors.

$$Pre \; \underset{\Downarrow}{\overset{\Uparrow}{\rightleftarrows}} \; Top \; \underset{\Xi}{\overset{\mathcal{O}}{\rightrightarrows}} \; Pos$$

The two on the left are covariant. They also form an ADJUNCTION, but we won't explain that until Chapter 4. The two on the right are contravariant. They are also NATURALLY ISOMORPHIC, and we explain that later in this chapter.

You should remember that a topological space need not be hausdorff. The separation properties T_0 and T_1 play a minor role here.

We look first at the two covariant functors $\Uparrow$ and $\Downarrow$ on the left.

Consider an arbitrary preset A. An upper section of A is a subset $U \subseteq A$ such that

$$\left. \begin{array}{r} a \in U \\ a \leq b \end{array} \right\} \Longrightarrow b \in U$$

for all $a, b \in A$. Let ΥA be the family of all upper sections of A. This is a topology on A, and is sometimes called the Alexandroff topology.

Let $\Uparrow A$ be the preset A viewed as a topological space, that is with ΥA as the carried topology. We think of $\Uparrow A$ as an upgrading of A. (It's getting above itself.) This gives the object assignment of one of the functors. The arrow assignment is more or less trivial.

For a topological space S with topology $\mathcal{O}S$, the specialization order of S is the comparison on S given by

$$r \leq s \iff (\forall U \in \mathcal{O}S)[r \in U \Longrightarrow s \in U] \iff r \in s^-$$

for $r, s \in S$. Here s^- is the closure of $\{s\}$. This specialization order is a pre-order on S. (You might like to check that S is T_0 precisely when $\leq$ is a partial ordering, and S is T_1 precisely when $\leq$ is equality.)

Let $\Downarrow S$ be the space S viewed as a preset, that is with its specialization order as its carried comparison. We think of $\Downarrow S$ as a downgrading of S. (It isn't making enough of its talents.) This gives the object assignment of the other functor. The arrow assignment is more or less trivial.

The Exercises 3.3.3, 3.3.4, and 3.3.5 fill in some of the missing details.

Next we look at the two contravariant functors $\mathcal{O}$ and Ξ on the right.

For each space S let $\mathcal{O}S$ be its topology viewed as a poset under inclusion. For each continuous map

$$T \xrightarrow{\ \phi\ } S$$

between spaces consider the inverse image function.

$$\mathcal{O}S \xrightarrow{\ \mathcal{O}(\phi) = \phi^{\leftarrow}\ } \mathcal{O}T$$

Almost trivially, this is monotone, and so gives us one of the functors.

Consider the 2-element set on the left

$$\mathbf{2} = \{0, 1\} \qquad \mathcal{O}\mathbf{2} = \{\emptyset, \{1\}, \mathbf{2}\}$$

together with the topology $\mathcal{O}\mathbf{2}$ on the right. This is Sierpiński space.

For a space S consider the set of **continuous characters** of S.

$$\Xi S = \mathbf{Top}[S, \mathbf{2}]$$

These are partially ordered pointwise, that is

$$p \leq q \iff (\forall s \in S)[p(s) \leq q(s)]$$

for $p, q \in \Xi S$. For each continuous map

$$T \xrightarrow{\ \phi\ } S$$

between space let

$$\Xi S \xrightarrow{\ \Xi(\phi)\ } \Xi T$$
$$p \longmapsto p \circ \phi$$

for $p \in \Xi S$. This gives us the other functor. Of course, there are a few things to be checked. These are dealt with by Exercises 3.3.6 and 3.3.7.

Exercises

3.3.3 (a) For a preset A, what is the specialization order of $\Uparrow A$?
(b) For a space S, show that $\mathcal{O}S \subseteq \Upsilon \Downarrow S$.

3.3.4 (a) Show that a monotone function

$$A \xrightarrow{\ f\ } B$$

between presets is continuous relative to the two Alexandroff topologies.
Show that $\Uparrow$ is a functor.
(b) Show that a continuous map

$$S \xrightarrow{\ \phi\ } T$$

between spaces is monotone relative to the two specialization orders.
Show that $\Downarrow$ is a functor.

3.3.5 Let

$$\theta : A \longrightarrow S$$

be a function from a pre-ordered set to a topological space.
Show that θ is monotone (relative to $\Downarrow S$) precisely when it is continuous (relative to ΥA).
Show there is a bijection between

$$\boldsymbol{Pre}[A, \Downarrow S] \qquad \boldsymbol{Top}[\Uparrow A, S]$$

for arbitrary A and S.

3.3.6 Show that for each continuous map

$$T \xrightarrow{\ \phi\ } S$$

between spaces, the function $\mathcal{O}(\phi)$ is monotone.
Show that $\mathcal{O}(\phi)$ passes across composition, and hence $\mathcal{O}$ is a functor.
Show that the function $\Xi(\phi)$ does convert continuous characters into continuous characters, and that $\Xi(\phi)$ is monotone.
Show that $\Xi(\phi)$ passes across composition, and hence Ξ is a functor.
Where have you seen some of these calculations before?

3.3.7 For an arbitrary space S and open set $U \in \mathcal{O}S$, let

$$\chi_S(U) : S \longrightarrow 2$$

be the characteristic function of U, that is

$$\chi_S(U)(s) = \begin{cases} 1 \text{ if } s \in U \\ 0 \text{ if } s \notin U \end{cases}$$

for $s \in S$.

Show that for each $U \in \mathcal{O}S$ the character $\chi_S(U)$ is continuous, and hence we have an assignment

$$\mathcal{O}S \xrightarrow{\;\chi_S\;} \Xi S$$

between the two posets.

Show that χ_S is an isomorphism of posets. (This is more than showing χ_S is a monotone bijection.)

3.3.3 Functors from products

In Section 2.5 we defined the notion of a product of two objects A and B in a category C. This consists of a wedge

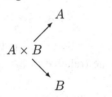

with certain properties. As we saw in Section 2.5 the object $A \times B$ is unique *only* up to a certain canonical isomorphism. What happens if we decide to change some of these selected objects and modify the projections accordingly?

Suppose the category C has all binary products. For each pair A, B of objects suppose we select, in some way or other, a product wedge for that pair. This choice could be haphazard, but it still produces a functor.

Let R be some fixed object of C. For each object A consider a product wedge, as on the right,

$$F = A \times R \qquad\qquad\quad \begin{array}{c} A \\ \nearrow p_A \\ A \times R \\ \searrow q_A \\ R \end{array}$$

together with the product object FA, as on the left. Thus we have an object assignment

$$A \longmapsto FA$$

on C. We show there is a corresponding arrow assignment

$$f \longmapsto F(f)$$

so that the pair of assignments forms an endo-functor on C.
   Let

$$A \xrightarrow{\ f\ } B$$

be an arrow of C. We have a diagram

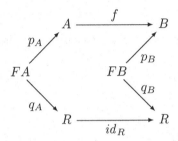

using the selected product wedges. The product condition gives an arrow

$$FA \xrightarrow{\ F(f)\ } FB$$

with certain properties. With a little bit of work we can check that this gives an endo-functor on C.

Exercises

3.3.8   Use the mediating property for product wedges to complete the details of the functorial product construction.
   Show also that the coproduct construction gives an endo-functor on the parent category.

3.3.9   If you are happy with the previous exercise, you can try this generalization. Let C be a category with all binary products, and consider the assignment

$$C \times C \longrightarrow C$$
$$(A_1, A_2) \longmapsto A_1 \times A_2$$

which attaches a product object to each pair of objects. Show that this fills out to a functor.

3.3.4 Comma category

In this block we use two functors to produce a new category from three old categories. This construction generalizes the two slice constructions of Example 1.3.6.

We start with three categories and two functors

$$U \xrightarrow{\ U\ } C \xleftarrow{\ L\ } L$$

where we think of U as the upper component and L as the lower component. Using these we produce a category

$$(U \downarrow L)$$

sometimes called a comma category. Each new object is a triple (conveniently written vertically)

$$
\begin{array}{ccc}
A_U & U A_U & A_U \in U \\
\big| & \big| & \\
\alpha & \alpha & \alpha \in C \\
\big\downarrow & \big\downarrow & \\
A_L & L A_L & A_L \in L
\end{array}
$$

formed using an upper object A_U from U, a lower object A_L from L, and a connecting central arrow α from C, as indicated. The new arrows

$$
\begin{array}{ccc}
U A_U & & U B_U \\
\big| & & \big| \\
\alpha & \xrightarrow{\ f\ } & \beta \\
\big\downarrow & & \big\downarrow \\
L A_L & & L B_L
\end{array}
$$

are formed using an arrow f_U from U and an arrow f_L from L such that the square

$$
\begin{array}{ccc}
A_U & \xrightarrow{\ f_U\ } & B_U \\
U A_U & \xrightarrow{\ U(f_U)\ } & U B_U \\
\big| & & \big| \\
\alpha & & \beta \\
\big\downarrow & & \big\downarrow \\
L A_L & \xrightarrow{\ L(f_L)\ } & L B_L \\
A_L & \xrightarrow{\ f_L\ } & B_L
\end{array}
$$

commutes. You should check that this does produce a category and generalizes the two slice constructions.

Exercises

3.3.10 Fill in the details of the construction of the comma category.

3.3.11 (a) What is $(U \downarrow L)$ when both U and L are the identity endo-functor on C?

(b) For an object S of C describe the slice categories $(C \downarrow S)$ and $(S \downarrow C)$ as comma categories.

3.3.12 For convenience let Com be the comma category $(U \downarrow L)$. Construct three forgetful functors

$$Com \longrightarrow U \qquad Com \longrightarrow C^{\downarrow} \qquad Com \longrightarrow L$$

using the arrow category in the central one.

3.3.5 Other examples

Functors appear almost everywhere in mathematics. Exercises 3.3.13 to 3.3.18 give a few more examples.

Exercises

3.3.13 For a group A let δA be the derived subgroup (generated by the commutators). In particular, $A/\delta A$ is an abelian group. Show that each of the two object assignments

$$A \longmapsto \delta A \qquad\qquad A \longmapsto A/\delta A$$

is part of a functor.

3.3.14 Consider a morphism between monoids.

$$S \xrightarrow{\ \phi\ } R$$

Using restriction of scalars we may view each (right) R-set A as an S-set. The S-action $\star$ is obtained from the R-action $\cdot$ by

$$a \star s = a \cdot \phi(s)$$

for each $a \in A$ and $s \in S$.

(a) Show that this construction does convert the R-set A into an S-set.

(b) Show that the construction produces a functor

$$Set\text{-}S \xleftarrow{\ \Phi\ } Set\text{-}R$$

which is trivial on objects *and* arrows.

(c) Try generalizing this construction using rings and modules.

3.3.15   You will have to think clearly to do this exercise.

We form a new large category $\boldsymbol{MON}$. Each object of $\boldsymbol{MON}$ is a category $\boldsymbol{Set}$-R for some monoid R. The arrows of $\boldsymbol{MON}$ are the functors between these categories. Show that the construction of Exercise 3.3.14 produces a contravariant functor

$$\boldsymbol{Mon} \longrightarrow \boldsymbol{MON}$$

from the small to the large.

3.3.16   Exercise 1.2.7 shows that the two categories $\boldsymbol{Set}_\perp$ and $\boldsymbol{Pfn}$ are 'essentially the same' category. Re-do that exercise to show there is an inverse pair of functors passing between the two categories.

3.3.17   Each preset S can be converted into a poset in a canonical fashion. We consider the relation $\sim$ on S given by

$$s_1 \sim s_2 \iff s_1 \leq s_2 \text{ and } s_2 \leq s_1$$

for $s_1, s_2 \in S$. Almost trivially, this is an equivalence relation on S, and is equality precisely when S is a poset.

Let $S/\!\!\sim$ be the corresponding set of blocks $[s]$ for $s \in S$, and partially order $S/\!\!\sim$ by

$$[s_1] \leq [s_2] \iff s_2 \leq s_2$$

for $s_1, s_2 \in S$.

(a) Show that this construction of a poset $S/\!\!\sim$ is well-defined, and show that the canonical function

$$S \longrightarrow S/\!\!\sim$$

is monotone.

(b) Show that

$$S \longmapsto S/\!\!\sim$$

is the object part of a functor $\boldsymbol{Pre} \longrightarrow \boldsymbol{Pos}$.

3.3.18   This exercise makes precise the notion of 'freely generated by' in appropriate circumstances. Later we look at a more general version of this construction.

Suppose we have two categories $\boldsymbol{Src}$ and $\boldsymbol{Trg}$ and a forgetful functor between them. It is customary not to give such a functor a name, but here it will help if it does have one.

You are allowed not to take the following too seriously.
Let

$$\boldsymbol{Src} \xleftarrow{\quad \iota \quad} \boldsymbol{Trg}$$

be the forgetful functor. (Eventually you can forget 'ι'!)

Suppose to each **Src**-object A we attach a **Trg**-object FA and an arrow

$$A \xrightarrow{\eta_A} (\iota \circ F)A$$

of **Src** with the following universal property.

For each **Src**-arrow

$$A \xrightarrow{\quad f \quad} \iota S$$

where S is a **Trg**-object, there is a unique **Trg**-arrow

$$FA \xrightarrow{\quad f^\sharp \quad} S$$

such that the triangle

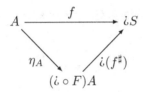

commutes in **Src**.

(a) Show that

$$A \longmapsto FA$$

is the object assignment of a functor $\boldsymbol{Src} \longrightarrow \boldsymbol{Trg}$.

(b) Show that for each **Src**-object A and **Trg**-object S the assignment

$$f \longmapsto f^\sharp$$
$$\boldsymbol{Src}[A, \iota S] \longrightarrow \boldsymbol{Trg}[FA, S]$$

is a bijection, and describe its inverse.

(c) Where have you seen this construction before?

3.4 Natural transformations defined

As we have seen, each arrow of a category compares two objects, and each functor compares two categories. Next we will see how each natural transformation compares two functors.

How might we compare two functors F and G? Surely we want them to pass between the same two categories

$$Src \underset{G}{\overset{F}{\rightrightarrows}} Trg$$

in the same direction. It also seems reasonable to insist that they have the same variance, either both are covariant or both are contravariant. Given these condition, how might we compare F and G?

Consider an arbitrary object A of *Src*. The two functors pass this to two objects FA and GA of *Trg*. We compare these objects in *Trg*. Thus we look for an arrow

$$FA \xrightarrow{\tau_A} GA$$

of *Trg*. We do this for each object A of *Src*.

3.4.1 Definition   (Preliminary) Given a parallel pair

$$Src \underset{G}{\overset{F}{\rightrightarrows}} Trg$$

of functors of the same variance, a natural transformation

$$F \xrightarrow{\ \tau\ } G$$

is a family of arrows of *Trg*

$$FA \xrightarrow{\tau_A} GA$$

indexed by the objects A of *Src*.   □

Notice that each component arrow τ_A passes in the same direction, from F to G in this case. Of course, there is more to a natural transformation than just an indexed family of arrows. The selected arrow τ_A is required to be natural for variation of A. This is where we have to take note of the common variance of F and G.

3.4.2 Definition (In full) Given a parallel pair

$$\mathbf{Src} \underset{G}{\overset{F}{\rightrightarrows}} \mathbf{Trg}$$

of functors of the same variance, a natural transformation

$$F \xrightarrow{\ \tau\ } G$$

is a family of arrows of **Trg**

$$FA \xrightarrow{\ \tau_A\ } GA$$

indexed by the objects A of **Src**, and such that for each arrow

$$A \xrightarrow{\ f\ } B$$

of **Src** the appropriate square in **Trg** commutes

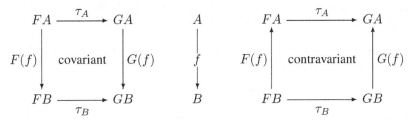

depending on the common variance of F and G. □

This is quite a short definition, but it has some subtleties. This will become clear as we look at various examples in the next section.

A natural transformation compares two functors. We refine the notion to make precise the idea of two functors being 'essentially the same'.

3.4.3 Definition A natural isomorphism between two functors F and G is a natural transformation

$$F \xrightarrow{\ \tau\ } G$$

such that for each source arrow A the selected arrow

$$FA \xrightarrow{\ \tau_A\ } GA$$

is an isomorphism in the target category. □

Sometimes two functors that are naturally isomorphic are said to be naturally equivalent.

Exercises

3.4.1 (a) Consider the small graph ($\downarrow$) as described in Example 1.3.5. We may view this as a very small category with two objects and three arrows. The two identity arrows have been omitted from the picture.

Show that the objects of $C^{\downarrow}$ are essentially the covariant functors

$$(\downarrow) \longrightarrow C$$

and the arrows of $C^{\downarrow}$ are the natural transformations between these functors.

(b) Show that each of the categories of Exercise 1.3.10 is the category of functors

$$\nabla \longrightarrow C$$

and natural transformations for some appropriate template category ∇.

(c) Can you see a generalization of this idea?

3.4.2 For an arbitrary poset S consider the category $\widehat{S}$ of presheaves on S, as defined in Example 1.4.1. Describe this as a category of functors and natural transformations.

3.4.3 Let R be a monoid viewed as a category. Exercise 3.2.3 located the functors of both variance.

$$R \longrightarrow Set$$

Now locate the natural transformations between these functors.

3.4.4 Let

$$F \xrightarrow{\ \ \tau\ \ } G$$

be a natural isomorphism between two functors. Suppose that for each source object A the arrow

$$FA \xrightarrow{\ \ \tau_A\ \ } GA$$

has an inverse in the target category.

$$FA \xleftarrow[\sigma_A]{\ \ \ \ \ \ } GA$$

Show that the family σ of arrows is a natural transformation.

3.5 Examples of natural transformations

In this section we look at several examples of natural transformations. Some of these build on earlier examples of functors. The exercises give further details and examples.

As a first example let's have a look at some natural transformations between hom-functors.

3.5.1 Example Let C be an arbitrary category, and let K and L be arbitrary objects of C. These give hom-functors

$$[K,-] \qquad [L,-]$$

from C to *Set*.

$$C \longrightarrow Set$$

As is customary, here we can omit the name of the parent category on the hom-functors. Consider an arbitrary arrow of C.

$$L \xrightarrow{\phi} K$$

For each object A of C we have an assignment

$$[K,A] \xrightarrow{\tau_A} [L,A]$$
$$l \longmapsto l \circ \phi$$

given by composition in C. We show these functions form a natural transformation between the two functors.

To do that we must consider an arbitrary arrow f of C, as on the left

and verify that the inner square commutes. To do that we take an arbitrary member

$$K \xrightarrow{l} A$$

of the top left hand corner of the square and track it both ways to the bottom right hand corner. We require that both paths give the same member of $[L, B]$. Thus we require

$$(f \circ l) \circ \phi = f \circ (l \circ \phi)$$

which is immediate. □

Work out the corresponding result for the contravariant hom-functors. Exercises 3.5.1 and 3.5.2 deal with this and a more involved version.

By inspecting the construction of the natural transformation τ of Example 3.5.1 we see that it is completely determined by one output

$$\phi = \tau_K(id_K)$$

of one component of τ. There is more to this.

3.5.2 Example Let C be an arbitrary category, let

$$C \xrightarrow{\ F\ } Set$$

be an arbitrary functor to *Set*. Let K be an arbitrary object of C. What can a natural transformations

$$[K, -] \longrightarrow F$$

look like? We show they are in bijective correspondence with the elements of the set FK.

(a) Consider first any element $k \in FK$. This gives a family of functions

$$
\begin{array}{c}
[K, A] \xrightarrow{\ \epsilon_A\ } FA \\
l \longmapsto F(l)(k)
\end{array}
$$

indexed by the objects A of C. This function ϵ_A is 'evaluation at k'.

We check that the family

$$[K, -] \xrightarrow{\ \epsilon\ } F$$

is a natural transformation. To do that we must consider an arbitrary arrow f of C, as on the left, and verify that the inner square commutes.

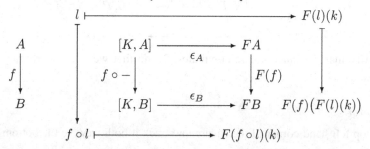

To do that we take an arbitrary member of the top left hand corner

$$K \xrightarrow{\quad l \quad} A$$

of the square and track both ways to the bottom right hand corner. We required that both paths give the same member of FB. Thus we require

$$F(f \circ l)(k) = F(f)\big(F(l)(k)\big)$$

for arbitrary f and l. But F is a covariant functor, so

$$F(f)\big(F(l)(k)\big) = \big(F(f) \circ F(l)\big)(k) = F(f \circ l)(k)$$

to give the required result.

(b) This gives us many examples of natural transformations from $[K, -]$ to F. Are there any more? In fact, we can show that every such natural transformation is determined by a unique element of FK.

Consider an arbitrary natural transformation

$$[K, -] \xrightarrow{\quad \tau \quad} F$$

look at the K-component

$$[K, K] \xrightarrow{\quad \tau_K \quad} FK$$

and set

$$k = \tau_K(id_K)$$

to produce $k \in FK$. We show that τ is 'evaluation at k'.

Consider an arbitrary object A of $\mathbf{C}$ and an arbitrary member

$$K \xrightarrow{\quad l \quad} A$$

of $[K, A]$. We show that

$$\tau_A(l) = F(l)(k)$$

holds. To do that we remember that the square

$$
\begin{array}{ccc}
[K, K] & \xrightarrow{\ \tau_K\ } & FK \\
{\scriptstyle l\,\circ\,-}\Big\downarrow & & \Big\downarrow{\scriptstyle F(l)} \\
[K, A] & \xrightarrow[\ \tau_A\ ]{} & FA
\end{array}
$$

commutes. By tracking the member id_K of the top left hand corner we obtain the required result. $\square$

Consider the natural transformation induced by $k \in FK$.

$$[K, -] \xrightarrow{\epsilon} F$$

When this is a natural isomorphism, we say the pair (K, k) is a **pointwise representation** of F. We say F is **representable** when it has at least one pointwise representation.

There is, of course, a contravariant version of this example, and in a way that is more important.

Let C be an arbitrary category. A **presheaf** on C is a contravariant *Set*-valued functor.

$$C \xrightarrow{F} Set$$

Such presheaves F and G are compared via natural transformations.

$$G \xrightarrow{\tau} F$$

These presheaves, as objects, and natural transformations, as arrows, form a category $\widehat{C}$, the Yoneda completion of C. Each object A of C gives a presheaf on C

$$\widehat{A} = C[-, A]$$

the contravariant hom-functor. These are the representable presheaves. Let A be a fixed object of C, and let F be a fixed presheaf on C. The basic Yoneda result characterizes the natural transformations

$$\widehat{A} \longrightarrow F$$

from the representable to the arbitrary. They are essentially the elements of the set FA. See Exercises 3.5.4 to 3.5.6. You might also want to have another look at Exercises 3.4.2 and 3.4.3.

In Block 3.3.1 we set up three endo-functors

$$\exists \quad | \quad \forall$$

on *Set*. There are several natural transformations associated with these.

3.5.3 Example   Let Id be the identity endo-functor on *Set*. Thus

$$Id\,A = A \qquad Id(f) = f$$

for each set A and function f. We set up two natural transformations

$$Id \xrightarrow{\eta^{\exists}} \exists \qquad\qquad Id \xrightarrow{\eta^{\forall}} \forall$$

using the two indicated covariant endo-functors on *Set*. Thus for each set A we require a pair a functions

$$A \xrightarrow{\eta_A^\exists} \mathcal{P}A \qquad\qquad A \xrightarrow{\eta_A^\forall} \mathcal{P}A$$

with appropriate properties. Notice that here we have omitted 'Id'. This should not cause too much confusion.

We must produce $\eta^\exists$ and $\eta^\forall$ so that for each function f the two squares

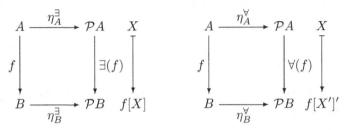

commute. For each $a \in A$ each of

$$\eta_A^\exists(a) \qquad \eta_A^\forall(a)$$

must be a certain subset of A. Given the other requirements, there isn't much choice. $\qquad\qquad\qquad\qquad\qquad\qquad\qquad\qquad\qquad\qquad\qquad$ □

The two natural transformations $\eta^\exists$ and $\eta^\forall$ of this last example don't look very interesting. However, in a more general setting they are quite important. We look at this in Chapter 6.

Let's now look at the contravariant power set functor. In the next example we set up a natural isomorphism which again doesn't look very exiting. However, the idea has many important refinements. It is the core of many representation results, especially when **2** is replaced by a more complicated structure.

3.5.4 Example The inverse image functor I on *Set* is contravariant, and is really a hom-functor in disguise. The set

$$\mathbf{2} = \{0, 1\}$$

induces a hom-functor $[-, \mathbf{2}]$ on *Set*. Thus we have two endo-functors

$$\mathbf{Set} \underset{[-, \mathbf{2}]}{\overset{I}{\rightrightarrows}} \mathbf{Set}$$

on *Set*, both of which are contravariant. We show that these two functors are naturally isomorphic.

To do that we recall that for any set A its subsets are in bijective correspondence with the characteristic functions on A. Thus, for each $X \in \mathcal{P}A$ we let

$$\chi_A(X) : A \longrightarrow 2$$

be given by

$$\chi_A(X)(a) = \begin{cases} 1 \text{ if } a \in X \\ 0 \text{ if } a \notin X \end{cases}$$

for $a \in A$. The assignment

$$\mathcal{P}A \xrightarrow{\chi_A} [A, 2]$$
$$X \longmapsto \chi_A(X)$$

is a bijection. We show this is natural for variation of A. To do that we must show that the inner square commutes

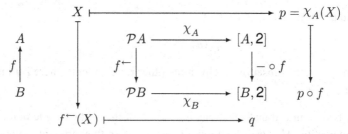

for an arbitrary arrow f, as on the left. Observe the contravariance here.

We track an arbitrary member $X \in \mathcal{P}A$ of the top left hand corner both ways to the bottom right hand corner. This gives us two members q and $p \circ f$ of $[B, 2]$. A calculation shows these are the same function. $\square$

Two compatible contravariant functors can be composed to produce a covariant functor. This often happens when we produce a 'representation' of an algebra. Let's look at a miniature version of that.

3.5.5 Example   The inverse image functor I on Set can be composed

$$\Pi = I \circ I$$

with itself to produce a covariant endo-functor on **Set**. Thus for each set A we have

$$\Pi A = \mathcal{P}^2 A$$

the second power set of A, the family of all collections of subsets of A. To describe the behaviour of Π on functions we fix some notation.

Each function f gives us two other functions

$$A \xrightarrow{\quad f \quad} B$$

$$\mathcal{P}A \xleftarrow{\quad \mathsf{I}(f) = f^{\leftarrow} \quad} \mathcal{P}A$$

$$\mathcal{P}^2 A \xrightarrow{\quad\quad\quad} \mathcal{P}^2 B$$
$$\Pi(f)$$

where the central one goes in the opposite direction. For the pair of sets

$$A \qquad\qquad\qquad B$$

we let

$$x \in A \quad X \in \mathcal{P}A \quad \mathcal{X} \in \mathcal{P}^2 A \qquad y \in B \quad Y \in \mathcal{P}B \quad \mathcal{Y} \in \mathcal{O}^2 B$$

be typical members of the indicated sets. We have

$$x \in \mathsf{I}(f)(Y) \iff f(x) \in Y$$

for each $Y \in \mathcal{P}B$ and $x \in A$. This gives

$$Y \in \Pi(f)(\mathcal{X}) \iff \mathsf{I}(f)(Y) \in \mathcal{X} \iff f^{\leftarrow}(Y) \in \mathcal{X}$$

for $\mathcal{X} \in \mathcal{P}^2 A$ and $Y \in \mathcal{P}B$.

For each set A let

$$A \xrightarrow{\quad \eta_A \quad} \mathcal{P}A$$

be the function given by

$$X \in \eta_A(x) \iff x \in X$$

for x, X as above. We show that

$$\mathrm{Id} \xrightarrow{\quad \eta \quad} \Pi$$

is a natural transformation.

We must show that the inner square commutes

for each function f as indicated on the left of the square. To do that, as usual, we take an arbitrary member x of the top left hand component and track it both ways to the bottom right hand component. Thus

$$\eta_B(f(x)) = \Pi(f)(\eta_A(x))$$

is the problem. This can be verified by a simple calculation.   □

As final example we look at one of the motivating 'natural' constructions and its 'unnatural' mate.

3.5.6 Example   Let K be a (commutative) field, and let **Vect**$_K$ be the category of vector spaces over K, or K-spaces for short. Each K-space V is an abelian group, written additively, and furnished with an action

$$K, V \longrightarrow V$$
$$r, a \longmapsto ra$$

satisfying the usual axioms. This is a left action, but since K is commutative the difference between left and right hardly matters.

These K-spaces are the objects of **Vect**$_K$, and the arrows are the corresponding linear transformations.

The field K is itself a K-space. Thus for an arbitrary K-space V we may form the hom-set

$$V^* = \mathbf{Vect}_K[V, K]$$

in **Vect**$_K$. It turns out that we can furnish V^* as a K-space to produce the **dual space** of V. In fact, $(\cdot)^*$ is a contravariant endo-functor on **Vect**$_K$. It is an enriched hom-functor.

We wish to investigate the interaction between a parent K-space V and its dual space V^*. To do that we fix some notation and terminology.

We let

$$r, s, t, \ldots \text{ range over scalars,} \quad \text{the members of } K$$
$$a, b, c, \ldots \text{ range over vectors,} \quad \text{the members of } V$$
$$\alpha, \beta, \gamma, \ldots \text{ range over characters, the members of } V^*$$

and we let

$$f, g, h, l, \ldots$$

range over linear transformations between K-spaces.

The elements of V^* are those functions

$$\alpha : V \longrightarrow K$$

such that

$$\alpha(0) = 0 \quad \alpha(a+b) = \alpha(a) + \alpha(b) \quad \alpha(ra) = r\alpha(a)$$

for all $a, b \in V$ and $r \in K$. We add these pointwise and this, with the obvious zero, furnishes V^* as an abelian group.

The action

$$K, V^* \longrightarrow V^*$$
$$r, \alpha \longmapsto r\alpha$$

is given by

$$(r\alpha)(a) = r(\alpha(a))$$

for $r \in K, \alpha \in V^*$, and $a \in V$. This converts V^* into a K-space.

Each finite dimensional K-space V is uniquely determined up to isomorphism. If V has dimension $n \geq 0$ then the isomorphisms

$$K^n \longrightarrow V$$

are in bijective correspondence with the bases of V. The crucial fact, which you should look up some time, is as follows.

Let V be a finite dimensional K-space. Then the dual space V^* is finite dimensional with the same dimension. In particular $V \cong V^*$.

This suggests a problem.

Let V be finite dimensional. There is at least one isomorphism

$$V \longrightarrow V^*$$

but is there a canonical one? To set up such an isomorphism we must first select a base for V, and then the resulting isomorphism is hardly canonical.

Now we come to what used to be the puzzling bit.

Each K-space V has a dual space V^* which itself has a dual space V^{**}. This is the **second dual** of V. We know that $(\cdot)^{**}$ is a covariant endo-functor on **Vect**$_K$ (because it is the composite of two contravariant endo-functors). Furthermore, it is easy to exhibit members of V^{**}.

For each $a \in V$ let

$$a^\wedge : V^* \longrightarrow K$$

be the function given by

$$a^\wedge(\alpha) = \alpha(a)$$

for $\alpha \in V^*$.

We can now check three facts.

(1) For each $a \in V$ the functions $a\hat{\ }$ is a member of V^{**}.

(2) For each K-space V the assignment

$$V \xrightarrow{\ (\cdot)\hat{\ }\ } V^{**}$$

is a linear transformation.

(3) The whole family of assignments $(\cdot)\hat{\ }$ is a natural transformation.

Thus for each finite dimensional K-space V the assignment of (2) is a canonical isomorphism, independent of any choice of base.   □

You can see what the puzzle was. Why is it that the second dual seems to have a 'natural' behaviour whereas the first dual doesn't?

We will meet many more functors and natural transformations. Some of these are quite complicated. The exercises give some hints of what can happen.

Exercises

3.5.1   Consider arbitrary objects K, L of an arbitrary category C. Show how a natural transformation

$$[-, L] \xrightarrow{\ \tau\ } [-, K]$$

can be induced by an arrow between K and L.

3.5.2   Let C be an arbitrary category and let

$$Q \xrightarrow{\ p\ } P$$

be an arbitrary arrow of C. Let

$$R \xrightarrow{\ s\ } S$$

be an arbitrary function (between sets). For each object A of C let

$$FA = \mathbf{Set}[C[A, P], R] \qquad GA = \mathbf{Set}[C[A, Q], S]$$

using hom-sets in the two different categories.

(a) Show that each of

$$A \longmapsto FA \qquad\qquad A \longmapsto GA$$

is the object assignment of a functor

$$C \longrightarrow \mathbf{Set}$$

and determine the variance of each.

(b) Use the arrow p and function s to produce a natural transformation $F \longrightarrow G$.

3.5.3 Complete the details of Example 3.5.2.

3.5.4 Consider the notion of a presheaf as defined just after Example 3.5.2. Where have you seen examples of this before?

3.5.5 Consider an arbitrary category C, and arbitrary presheaf F on C, an arbitrary object A of C, and an arbitrary element $a \in FA$.

For each set X consider the following assignment.

$$C[X, A] \xrightarrow{\overline{a}_X} FX$$
$$k \longmapsto F(k)(a)$$

Check that this is a function. In other words, show that the output does live in FX. Show that the whole family $\overline{a}$ is a natural transformation.

3.5.6 Continuing with the notation of Exercise 3.5.5, consider an arbitrary natural transformation

$$\widehat{A} \xrightarrow{\tau} F$$

and let $a = \tau_A(id_A)$. Check that $a \in FA$, and show that $\tau = \overline{a}$.

3.5.7 Describe the natural transformations $\eta^{\exists}$ and $\eta^{\forall}$ of Example 3.5.3.

3.5.8 Complete the calculation of Example 3.5.4.

3.5.9 (a) Do the calculation required to complete Example 3.5.5.

(b) By Example 3.5.4 the inverse image functor I is naturally isomorphic to the hom-functor $[-, 2]$. Thus Π is naturally isomorphic to the endo-functor with

$$A \longmapsto [[A, 2], 2]$$

as the object assignment. Write down the arrow assignment and re-do Example 3.5.5 for this functor.

(c) Which version do you think is easier to understand?

3.5.10 Consider Example 3.5.6

(a) Write down all the axioms needed to set up $\textbf{Vect}_K$. (The axioms for a field, for an additive abelian group, for an action, and for a linear transformation.) It is instructive not to overload the notation. In other words, distinguish between the various additions, and use a different visible infix for each multiplication.

(b) Verify that the dual space V^* of a K-space is itself a K-space. (You may now go back to the standard, overloaded, notation.)

(c) Show that $(\cdot)^*$ is a contravariant endo-functor on $\textbf{Vect}_K$. In particular, you must decide how $(\cdot)^*$ behaves on arrows of $\textbf{Vect}_K$.

3.5.11   Continuing with Example 3.5.6, verify the three facts (1), (2), and (3).

3.5.12   Show that the functors $\mathcal{O}$ and Ξ of Block 3.3.2 are naturally isomorphic.

3.5.13   Consider the functors arising from the product construction, as described in Block 3.3.3 and Exercises 3.3.8 and 3.3.9. Show that the projection arrows form natural transformations.

3.5.14   Let C be a category with all binary products. Let R, S be two objects and let

$$F = - \times R \qquad G = - \times S$$

to obtain two endo-functors on C. Show that each arrow

$$R \xrightarrow{\phi} S$$

of C induces a natural transformation

$$F \xrightarrow{\phi_\bullet} G$$

between these functors. This involves some serious diagram chasing.

3.5.15   Let C be a category with all binary products and coproducts, and let A, B, C be three arbitrary objects of C. Let

$$L = A \times C + B \times C \qquad R = (A + B) \times C$$

to form two more objects.

Show that by fixing two of A, B, C, each of L and R is an endo-functor of C, and there is a natural transformation $L \longrightarrow R$.

If you are brave you might try the 3-placed version of this, that is do not fix two of A, B, C.

3.5.16   Recall the difference between a monoid and a semigroup. (A semigroup need not have a unit.) Given a semigroup A let

$$FA = A \cup \{\omega\}$$

where ω is a new element not in A. Let

$$A \xrightarrow{\iota} FA$$

be the insertion. Let $\star$ be the operation on FA given by

$$a \star b = ab \qquad a \star \omega = a = \omega \star a \qquad \omega \star \omega = \omega$$

for all $a, b \in A$.

(a) Show that $(FA, \star, \omega)$ is a monoid.

(b) Show that ι is a semigroup morphism.

(c) Show that for each semigroup morphism

$$A \xrightarrow{\ f\ } B$$

to a monoid B, there is a commuting triangle

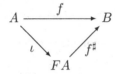

for some unique monoid morphism $f^\sharp$.

(d) Show that F fills out to a functor. You will have to sort out its source and target.

(e) Show that ι is natural for variation of A. You will have to insert a couple of trivial functors.

(f) What happens if A already has a unit?

3.5.17 For an arbitrary set A think of A as an alphabet. Let FA be the set of all words on A, finite lists

$$\mathsf{a} = [a_1, \ldots, a_l]$$

for $a_1, \ldots, a_l \in A$. The empty word, with $l = 0$, is allowed.

(a) Show that FA is a monoid under concatenation.

(b) Show that F fills out to a functor. Make sure you write down the source and target.

(c) Show that the assignment

$$\begin{array}{ccc} A & \longrightarrow & FA \\ a & \longmapsto & [a] \end{array}$$

is a natural transformation. You will have to sort out the two functors it passes between.

(d) Show that FA is the free monoid on the sets A in a sense that you should make precise.

(e) What happens if A already carries a monoid structure?

3.5.18   Consider the two functors

$$A \longmapsto \delta A \qquad\qquad A \longmapsto A/\delta A$$

of Exercise 3.3.13 (where here only the object assignments are given).

(a) Show the canonical embedding ι and the canonical quotient η

$$\delta A \xrightarrow{\ \iota\ } A \qquad\qquad A \xrightarrow{\ \eta\ } A/\delta A$$

are natural for variation of A. You must describe explicitly the source and target for each functor.

(b) Show that for each morphism from an arbitrary group A

$$A \xrightarrow{\ f\ } B$$

to an abelian group B, there is a unique morphism

$$A/\delta A \xrightarrow{\ f^\sharp\ } B$$

such that

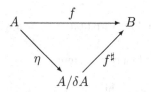

commutes.

3.5.19   Consider the 'freely generated by' construction of Exercise 3.3.18. Show that the family η of arrows is a natural transformation.

3.5.20   Let ∇ be an arbitrary category and think of this as a template. Let C be an arbitrary category. These combine to give another category C^∇. The objects of C^∇ are the covariant functors

$$\nabla \longrightarrow C$$

and the arrows are the natural transformations between these functors.

Show that this does give a category. The problem is to produce an appropriate method of composing natural transformations.

3.5.21   The composition used in Exercise 3.5.20 is know as the vertical composition of natural transformations. There is also a horizontal composition. (Don't ask what is vertical or horizontal about the two constructions.)

Consider three categories A, B, C, four functors F, G, K, L, and two natural transformations λ, ρ, as shown.

$$
\begin{array}{c}
A \\
F \Big| \!-\!\lambda\!\rightarrow\! \Big| G \\
B \\
K \Big| \!-\!\rho\!\rightarrow\! \Big| L \\
C
\end{array}
$$

Show that for each object A of A the following square commutes.

$$
\begin{array}{ccc}
(K \circ F)A & \xrightarrow{\;\rho_{FA}\;} & (L \circ F)A \\
\Big\downarrow {\scriptstyle K(\lambda_A)} & & \Big\downarrow {\scriptstyle L(\lambda_A)} \\
(K \circ G)A & \xrightarrow[\;\rho_{GA}\;]{} & (L \circ G)A
\end{array}
$$

Let

$$
(K \circ F)A \xrightarrow{\;(\rho \star \lambda)_A\;} (L \circ G)A
$$

be the diagonal of this square. Show that this family $(\rho \star \lambda)_\bullet$ is natural.

This gives the horizontal composite of

$$
(K \circ F) \xrightarrow{\;\rho \star \lambda\;} (L \circ G)
$$

of the two natural transformations between the composite functors.

3.5.22 Consider three categories A, B, C, six functors F, G, H, K, L, M, and four natural transformations $\lambda, \mu, \rho, \sigma$, as shown.

$$
\begin{array}{c}
A \\
F \Big| \!-\!\lambda\!\rightarrow\! \overset{\big|}{\underset{\blacktriangledown}{G}} \!-\!\mu\!\rightarrow\! \Big| H \\
B \\
K \Big| \!-\!\rho\!\rightarrow\! \overset{\big|}{\underset{\blacktriangledown}{L}} \!-\!\sigma\!\rightarrow\! \Big| M \\
C
\end{array}
$$

Using vertical and horizontal composition (as in Exercises 3.5.20 and 3.5.21) show that

$$
(\sigma \star \mu) \circ (\rho \star \lambda) = (\sigma \circ \rho) \star (\mu \circ \lambda)
$$

holds.

4

Limits and colimits in general

In Chapter 2, Sections 2.3 to 2.7 we looked at some simple examples of limits and colimits. These are brought together in Table 2.1 which is repeated here as Table 4.1. In this chapter we generalize the idea.

Before we begin the details it is useful to outline the five steps we go through together with the associated notions for each step. After that we look at each step in more detail.

Template

This is the shape ∇ that a particular kind of diagram can have. It is a picture consisting of nodes (blobs) and edges (arrows). The central column of Table 4.1 lists a few of the simpler templates. Technically, a template is often a directed graph or more generally a category.

Diagram

This is an instantiation of a particular template ∇ in a category C. Each node of ∇ is instantiated with an object of C, and each edge is instantiated with an arrow of C. There are some obvious source and target restrictions that must be met, and the diagram may require that some cells commute. Thus we sometimes use a category as a template.

Posed problem

Each diagram in a category C poses two problems, the left (blunt end) problem and the right (sharp end) problem. We never actually say what the problem is (which is perhaps the reason why it is rarely mentioned) but we do say what a solution is. The idea is to find a 'best' solution.

Solution

A solution for a diagram in C is a nominated object X of C together with a collection of arrows. For a left solution all arrows start from X, and such a

Table 4.1 *Some simple limits and colimits – a repeat of Table 2.1*

	Limit	Template	Colimit
(1)	final object		initial object
(2)	binary product		binary coproduct
(3)	equalizer		co-equalizer
(4)	pullback		
(5)			pushout

gadget is often called a cone. For a right solution all arrows finish at X, and such a gadget is often called a co-cone. For both kinds of solutions the arrows must make various triangles commute.

Universal solution

A universal solution is a particular solution through which each solution (of that handedness) must pass via a unique mediating arrow. A limit is a universal left solution. A colimit is a universal right solution.

We now begin to look at each these notions in more detail.

4.1 Template and diagram – a first pass

Roughly speaking a template is a collection of nodes, each drawn as a •, and a collection of edges, each drawn as an arrow. Each edge passes from a particular node (its source) to a particular node (its target). In other words, a template is a directed graph. There may also be some commuting conditions on the edges, in which case the template is a category. We usually draw the edges as pointing from left to right.

We instantiate the template in a category C to produce a diagram in C. We replace each node by an object of C and we replace each edge by an arrow of C. We respect the source and target conditions and any commuting conditions that the template requires. We look for the left (blunt end) solutions or the right (sharp end) solutions. In particular, we look for a universal solution on the appropriate side, to obtain a limit (universal left solution) or a colimit (universal right solution).

Table 4.1 gives a few small templates. Let's look at a few more examples that are not so simple.

4.1.1 Examples   (1) Suppose we have a collection of nodes with no edges. It is convenient of think of this collection arranged vertically.

There may be infinitely many of these, finitely many, or none at all.

A limit for a corresponding diagram is a product and a colimit for a corresponding diagram is a coproduct (sometimes called a sum). We have already seen the case where there are zero, one, or just two nodes.

(2) Suppose we have a collection of nodes arranged in a line with an edge between adjacent nodes.

If there are only finitely many nodes then the posed problem isn't interesting (since any associated diagram has a left-most object and a right-most object). Thus we may as well suppose there are infinitely many nodes. We use the integers as nodes. This gives us three (or perhaps four) different templates.

The template may have a left-most node and go off to the right

$$0 \longrightarrow 1 \longrightarrow 2 \longrightarrow 3 \longrightarrow 4 \longrightarrow \cdots$$

in which case it is the colimit (right universal solution) that is interesting.

The template may have a right-most node and go off to the left

$$\cdots \longrightarrow 4 \longrightarrow 3 \longrightarrow 2 \longrightarrow 1 \longrightarrow 0$$

in which case it is the limit (left universal solution) that is interesting.

Notice that we have again used the natural numbers to label the nodes. The template may go off to the left and the right

$$\cdots \longrightarrow -3 \longrightarrow -2 \longrightarrow -1 \longrightarrow 0 \longrightarrow 1 \longrightarrow 2 \longrightarrow 3 \longrightarrow \cdots$$

$$\cdots \longrightarrow 3 \longrightarrow 2 \longrightarrow 1 \longrightarrow 0 \longrightarrow -1 \longrightarrow -2 \longrightarrow -3 \longrightarrow \cdots$$

in which case we may use the integers to label the nodes in one of two ways. Both these are useful in different circumstances. For a diagram of this shape both the limit and the colimit may be interesting.

(3) The template may be a collection of zig-zags

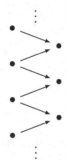

and may be finite or infinite. Even the finite case (with at least four nodes) leads to interesting solutions.

(4) There are more complicated examples.

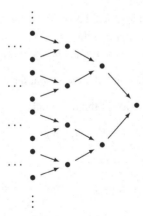

This is a tree which grows to the left and keeps growing for ever, but with the edges pointing to the right. A limit is something that is put out on the far left. What can that be?

Table 4.2 *A more exotic template*

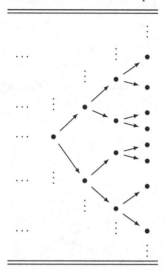

(5) Consider the template in Table 4.2. The nodes are arranged in vertical layers each of which is infinite both ways, and there are infinitely many layers progressing leftwards. Each edge passes between one layer and the next. Think of going through this graph from right to left, moving backwards along the edges. Intuitively something is being approached out on the far left. The notion of a limit makes this precise.

This template stops at the vertical layer on the right. We could also continue the same pattern moving off to the right. This would not change the limit (at the far left) but could have a dramatic impact on the colimit (at the far right).   □

Let's now try to make the idea of these examples precise.

4.1.2 Definition   A template (of the first kind)

$$\nabla = (\mathbb{I}, \mathbb{E})$$

is a directed graph consisting of

nodes $i, j, k, \ldots$ in $\mathbb{I}$ edges $e, f, g, \ldots$ in $\mathbb{E}$

where each edge

$$i \xrightarrow{\quad e \quad} j$$

has a nominated source and target, each of which is a node.   □

This version says nothing about any commuting conditions in the template. We look at that in the next section. However, notice that all the templates of Table 4.1 and Examples 4.1.1 do match this definition.

4.1.3 Definition   Let $\nabla = (\mathbb{I}, \mathbb{E})$ be a directed graph viewed as a template of the first kind. Let C be a category. A ∇-**diagram** in C is

- an $\mathbb{I}$-indexed family of objects of C $\qquad\qquad A = \big(A(i) \,|\, i \in \mathbb{I}\big)$
- an $\mathbb{E}$-indexed family of arrows of C $\qquad\qquad \mathcal{A} = \big(A(e) \,|\, e \in \mathbb{E}\big)$

where each edge, as on the left, produces an arrow, as on the right

$$i \xrightarrow{\;\;e\;\;} j \qquad\qquad A(i) \xrightarrow{\;\;A(e)\;\;} A(j)$$

with indicated source and target restrictions.   □

In other words, this is just a 'functor' from $\nabla = (\mathbb{I}, \mathbb{E})$ but with any commuting conditions ignored.

There aren't many exercises concerned solely with templates, but the following construction should be looked at.

Consider the notion of a directed graph as given in Definition 4.1.2. Such a gadget consists of nodes and edges with two source and target assignments. This looks a bit like the notion of a category. Each category is a directed graph, but the converse does not hold. A directed graph has no notion of composition of edges, and no notion of identity edges. However, there is a construction that converts each directed graph into a category.

4.1.4 Definition   Let ∇ be a directed graph.
For each $l \in \mathbb{N}$ a **path** through ∇ of length l is a list of l edges

$$i(0) \xrightarrow{\;e(1)\;} i(1) \xrightarrow{\;e(2)\;} i(2) \xrightarrow{\qquad} \cdots\cdots \xrightarrow{\;e(l)\;} i(l)$$

where the target of each edge is the source of the next one. A path of length 1 is just an edge. A path of length 0 is just a node.

We create a category $Pth(\nabla)$, the **category of paths** through ∇.
The objects of $Pth(\nabla)$ are the nodes of ∇.
The arrows of $Pth(\nabla)$ are the paths through ∇.
Given two paths

$$i(0) \longrightarrow i(1) \longrightarrow \cdots \longrightarrow i(l) \qquad j(0) \longrightarrow j(1) \longrightarrow \cdots \longrightarrow j(m)$$

with $i(l) = j(0)$ the composite path is

$$i(0) \longrightarrow i(1) \longrightarrow \cdots \longrightarrow i(l) = j(0) \longrightarrow j(1) \longrightarrow \cdots \longrightarrow j(m)$$

formed by sticking one path after the other.   □

There is something to be checked here, and this is not entirely trivial.

4.1.1   Show that the construction $Pth(\cdot)$ does produce a category.

Make sure you verify all the required properties. This is an example where the required identity properties are not immediately obvious.

4.1.2   Consider the following three different graphs.

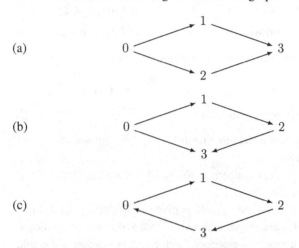

(a)

(b)

(c)

In each case describe the generated category of paths.

4.1.3   Suppose the graph you start from is already a category. Do you merely reconstruct the category?

4.1.4   Let ∇ be a directed graph viewed as a template. Let A be a ∇-diagram in some category C.

Show that A extends uniquely to a functor $Pth(\nabla) \longrightarrow C$.

Show that each functor $Pth(\nabla) \longrightarrow C$ is the unique extension of some ∇-diagram in C.

4.2 Functor categories

In this section we replace the directed graph $(\mathbb{I}, \mathbb{E})$ by a category ∇. We think of ∇ as an indexing gadget, and to emphasize this we refer to the objects $i, j, k, \ldots$ of ∇ as nodes and its arrows $e, f, g, \ldots$ as edges.

The following notion generalizes that of Example 1.3.5 and Exercise 1.3.10. You may want to look at those before you continue. You should make sure you understand they are particular cases of the following.

4.2.1 Definition   Let ∇ be an arbitrary category viewed as a template. Let C be an arbitrary category. These combine to produce the category

$$C^\nabla$$

of ∇-diagrams in C.

Each object of C^∇ is a functor

$$\nabla \xrightarrow{\;\;A\;\;} C$$

from the template to C.

Given two such functors (objects of C^∇)

$$\nabla \underset{B}{\overset{A}{\rightrightarrows}} C$$

an arrow of C^∇ from A to B

$$A \xrightarrow{\;\;\sigma\;\;} B$$

is a natural transformation between the functors.

Given three such functors and two such natural transformations

$$A\Bigg\downarrow \xrightarrow{\;\sigma\;} B\Bigg\downarrow \xrightarrow{\;\tau\;} C\Bigg\downarrow$$

the composite

$$A\Bigg\downarrow \xrightarrow{\;\tau\circ\sigma\;} C\Bigg\downarrow$$

is given by

$$A(i) \xrightarrow{\;(\tau\circ\sigma)_i = \tau_i\circ\sigma_i\;} C(i)$$

for each index i. □

There is something to prove here. We must show that the composite $\tau\circ\sigma$ of two natural transformations is a natural transformation, and that this composition is associative. The proofs are straightforward, but you should go through them. Because if you don't you know what will happen, don't you!

As the terminology of Definition 4.1.2 suggests, for each category ∇ viewed as a template, and each category C, a ∇-diagram in C is merely a functor A from ∇ to C. In other words, such a diagram is

a family of objects $A(i)$ of C a family of arrows $A(e)$ of C

indexed by the

nodes edges

of ∇, respectively. As with a diagram over a directed graph we require that each edge

$$i \xrightarrow{\ e\ } j$$

of ∇ produces an arrow

$$A(i) \xrightarrow{\ A(e)\ } A(j)$$

of C. We now also require that for each pair

$$i \xrightarrow{\ e\ } j \xrightarrow{\ f\ } k$$

of composable edges of ∇, the induced triangle in C

$$A(i) \xrightarrow{\ A(f \circ e)\ } A(k)$$
$$A(e) \searrow \qquad \nearrow A(f)$$
$$A(j)$$

commutes. Finally, we now also require that

$$A(id_i) = id_{A(i)}$$

for each index i.

In many cases we don't use an arbitrary category as a template. We use a partially ordered set, or occasionally a pre-ordered set.

Let $\mathbb{I}$ be a pre-ordered set, let

$$i, j, k, \ldots$$

range over $\mathbb{I}$ and think of these as nodes. We may view $\mathbb{I}$ as a category in one of two ways. For each pair of nodes i, j there is at most one edge

$$i \xrightarrow{\ (j,i)\ } j$$

from i to j. There is such an edge precisely when there is a comparison between i and j. We orientate these edges in one of two ways.

Upwards   Downwards

$i \leq j$ $i \xrightarrow{(j,i)} j$ $j \leq i$

Depending on which view we take, such an arrow always points upwards or always points downwards in the pre-ordered set.

4.2.2 Definition   A pre-ordered set is directed or upwards directed if for each pair i, j of nodes there is at least one node k with $i \leq k$ and $j \leq k$. $\square$

Depending on the circumstances sometimes we want a pre-ordered diagram that is directed to the right (directed to the sharp end). In that case we index the diagram by a directed pre-order with its edges pointing upwards. Sometimes we want a pre-ordered diagram that is directed to the left (directed to the blunt end). In that case we index the diagram by a directed pre-order with its edges pointing downwards. We could achieve the same effect using a *downwards* directed pre-order, but that rarely seems to be used.

Notice that when we use a pre-ordered set at a template we lie it on its side, so that 'upwards' means 'towards the right'.

To conclude this section let's take a closer look at the way edges in a preset are labelled. Since there is at most one edge between two indexes we can use these to label the edge.

$$i \xrightarrow{(j,i)} j$$

At first sight the two components of the edge seem to be the wrong way round. But consider what happens when we compose two such edges. The common node should disappear.

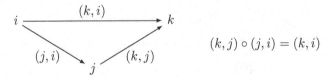

$$(k,j) \circ (j,i) = (k,i)$$

This convention is neater.

Exercises

Throughout these exercises ∇ and C are an arbitrary pair of categories.

4.2.1   Show that the construction of C^{∇} does produce a category.

4.2.2 For each C-object X let ΔX be the ∇-diagram with X at each node and id_X at each edge. Show that $X \longmapsto \Delta X$ is the object assignment of a functor

$$C \longrightarrow C^\nabla$$

from C to C^∇. This is the diagonal functor.

4.2.3 Describe typical arrows

$$\Delta X \longrightarrow A \qquad\qquad A \longrightarrow \Delta X$$

where X is a C-object and A is a ∇-diagram.

4.3 Problem and solution

Let ∇ be a template, let C be a category, and let A be a ∇-diagram in C. This diagram A is a collection of objects and arrows given by the shape ∇. This diagram poses two problems in C, the blunt end problem and the sharp end problem. When we draw the diagram we usually let the arrows point from left to right, so it is more common to speak of the left problem and the right problem.

The following definition is two definitions in one given in parallel.

4.3.1 Definition Let ∇ be a template, and let A be a ∇-diagram in a category C. A

| left | right |

solution is a C-object X together with a family of arrows

$$X \xrightarrow{\;\alpha(i)\;} A(i) \qquad\qquad A(i) \xrightarrow{\;\alpha(i)\;} X$$

indexed by the nodes of ∇, such that for each edge e of ∇

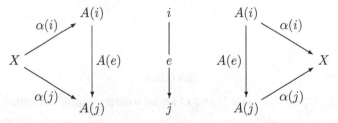

the induced C-triangle commutes.

A diagram may have many different left solutions or it may have none. It may have many right solutions or it may have none. In general there is little or no relationship between left solutions and right solutions. In the next section, we make precise the notion of a 'best' left solution or a 'best' right solution, but we don't need to worry about that just yet.

4.3.2 Example Suppose the category C is a poset with its arrows pointing upwards (to the right). One kind of diagram in C is simply a subset S with the induced comparisons. Then a

left   right

solution for S is merely a

lower   upper

bound of S in the poset. The orientations here have got a bit twisted, but that is just an historical accident. □

Is there something missing in Definition 4.3.1? Suppose the template imposes some commuting conditions on the diagram. Shouldn't those conditions be observed in the corresponding solutions? They are!

Suppose the template ∇ has composible edges

$$i \xrightarrow{\quad e \quad} j \xrightarrow{\quad f \quad} k$$

with

$$i \xrightarrow{\quad g = f \circ e \quad} k$$

as the composite edge. For a left solution we certainly require that the three triangles commute.

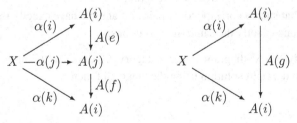

But the notion of a diagram requires

$$A(g) = A(f) \circ A(e)$$

so that the fact that the two triangles on the left commute ensures that the triangle on the right also commutes. For this reason sometimes only a 'generating part' of a template is used.

4.3.3 Example   Suppose we use the integers $\mathbb{Z}$

$$\cdots \quad -3 \quad -2 \quad -1 \quad 0 \quad 1 \quad 2 \quad 3 \quad \cdots$$

as a poset template with its edges pointing upwards. We have an edge

$$n \xrightarrow{\quad (m,n) \quad} m$$

for all $n \leq m$ in $\mathbb{Z}$. In fact

$$(m,n) = (m, m-1) \circ (m-1, m-2) \circ \cdots \circ (n+2, n+1) \circ (n+1, n)$$

using the 1-step edges. Any $\mathbb{Z}$-diagram requires

$$A(m,n) = A(m, m-1) \circ A(m-1, m-2) \circ \cdots \circ A(n+2, n+1) \circ A(n+1, n)$$

so, in practice, we often describe just the 1-step arrows

$$A(n+1, n)$$

for $n \in \mathbb{Z}$.   □

On the whole we need not worry too much about the difference between a diagram indexed by a directed graph and one indexed by a category. In fact, as we have seen, each graph ∇ has an associated path category $Pth(\nabla)$ and these produce 'equivalent' diagrams in any category. These two diagrams have the same solutions.

Exercises

4.3.1   Let ∇ be a directed graph and let C be an arbitrary category. Recall that each ∇-diagram in C extends uniquely to a $Pth(\nabla)$-diagram, and each $Pth(\nabla)$-diagram arises from a ∇-diagram

Show that such a corresponding pair of diagrams have exactly the same left solutions and exactly the same right solutions.

4.3.2   Given a ∇-diagram A in a category C, describe the notions of a left solution and a right solution using the diagonal functor.

4.4 Universal solution

Each diagram in a category poses two problems, the left problem and the right problem. Each of these problems may have many solutions, and in general there is no relationship between the left solutions and the right solutions. We now look for a 'best' solution on either side. On any particular side such a solution need not exists, but if there is one, then all the 'best' solutions are canonically isomorphic.

A universal solution of a diagram is a particular solution that is as economical as possible, in that it is as 'near' to the diagram as possible. Here is the formal definition. As usual, it is two definitions in one.

4.4.1 Definition   Let ∇ be a template and let A be a ∇-diagram in a category C. A

| left universal solution | right universal solution |

usually called a

| limit | colimit |

for A is a particular solution

$$S \xrightarrow{\ \sigma(i)\ } A(i) \qquad\qquad A(i) \xrightarrow{\ \sigma(i)\ } S$$

such that for each solution

$$X \xrightarrow{\ \alpha(i)\ } A(i) \qquad\qquad A(i) \xrightarrow{\ \alpha(i)\ } X$$

there is a *unique* arrow

$$X \xrightarrow{\ \mu\ } S \qquad\qquad S \xrightarrow{\ \mu\ } X$$

such that for each node i the triangle

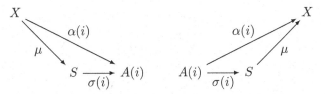

commutes. We call μ the mediator.   □

A limit or a colimit of a diagram need not exists. A diagram can have one without the other. However, once we have a universal solution (left or right) we can obtain all other solutions of the same handedness. We simply weaken the universal solution by an arrow to (for left) or from (for right) the carrying object, the apex of the universal solution.

Limits and colimits of diagrams (when they exist) are essentially unique. To prove that for limits we make a preliminary observation.

4.4.2 Lemma   *Let ∇ be a template, let A be a diagram in a category C, and suppose*

$$S \xrightarrow{\sigma(i)} A(i)$$

is a limit of that diagram indexed by the nodes i of ∇. Then this family of arrows is collection-wise monic. That is, if

$$X \underset{\psi}{\overset{\theta}{\rightrightarrows}} S$$

is a parallel pair of arrows with

$$\sigma(i) \circ \theta = \sigma(i) \circ \psi$$

for each node i, then $\theta = \psi$.

Proof   Consider any such parallel pair θ, ψ of arrows. Let

$$X \xrightarrow{\alpha(i)} A(i)$$

be the arrow given by

$$\alpha(i) = \sigma(i) \circ \theta = \sigma(i) \circ \psi$$

for node i. Consider any edge

$$i \xrightarrow{\quad e \quad} j$$

of ∇. By passing through S we see that the triangle

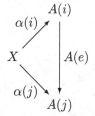

commutes, and hence we have a left solution of the diagram. Since S is the apex of a universal left solution there is a unique arrow

$$X \xrightarrow{\mu} S$$

such that

$$\alpha(i) = \sigma(i) \circ \mu$$

for each node i. Since both θ and ψ do this, we have $\theta = \psi$. $\square$

This result has a simple consequence.

4.4.3 Corollary   *Let ∇ be a template, let A be a diagram in a category C, and suppose*

$$S \xrightarrow{\ \sigma(i)\ } A(i)$$

is a limit of that diagram indexed by the nodes i of ∇. Suppose also that

$$S \xrightarrow{\ \epsilon\ } S$$

is an endo-arrow of S such that

$$\sigma(i) \circ \epsilon = \sigma(i)$$

for each node i. Then $\epsilon = id_S$.

Proof   We apply Lemma 4.4.2 to the pair $\theta = \epsilon$ and $\psi = id_S$. $\square$

So far we have been careful to speak of *a* limit of a diagram. We can now show that we needn't be so cautious.

4.4.4 Theorem   *Let ∇ be a template, let A be a diagram in a category C, and suppose each of*

$$S \xrightarrow{\ \sigma(i)\ } A(i) \qquad\qquad T \xrightarrow{\ \tau(i)\ } A(i)$$

is a limit of that diagram indexed by the nodes i of ∇. Then there is a unique arrow

$$T \xrightarrow{\ \tau\ } S$$

such that

$$\tau(i) = \sigma(i) \circ \tau$$

for each node i. Furthermore, τ is an isomorphism.

Proof   Since the arrows $\sigma(\cdot)$ form a limit, and the arrows $\tau(\cdot)$ form a left solution, there is a unique arrow τ, with indicated type, such that

$$\tau(i) = \sigma(i) \circ \tau$$

for each node i. This τ is just the mediator. By symmetry, there is a unique arrow

$$S \xrightarrow{\ \sigma\ } T$$

such that

$$\sigma(i) = \tau(i) \circ \sigma$$

for each node i.

Now consider the endo-arrow

$$\epsilon = \tau \circ \sigma$$

of S. For each node i we have

$$\sigma(i) \circ \epsilon = \sigma(i) \circ \tau \circ \sigma = \tau(i) \circ \sigma = \sigma(i)$$

and hence $\epsilon = id_S$ by Corollary 4.4.3. By symmetry we have

$$\tau \circ \sigma = id_S \qquad \sigma \circ \tau = id_T$$

hence these mediators σ and τ are an inverse pair of isomorphisms.   □

This shows that if a diagram has a limit then that limit is essentially unique. Thus we may speak of *the* limit of a diagram (when this does exist). There is a similar result for colimits with the same proof but where the arrows point the other way.

In the remaining three sections of this chapter we gather together a random collection of examples to show how limits and colimits can be calculated in appropriate circumstances.

Exercises

4.4.1   Find a simple example of a category where a diagram has two distinct limits.

4.4.2   Show that in the familiar categories (Set, Grp, Rng, ...) limits and colimits are not absolutely unique, only unique up to a canonical isomorphism.

4.4.3   State and prove the right hand version of each of the three results of this section.

4.5 A geometric limit and colimit

In this section we first look at a geometric example involving the circle group and topological spaces. After that there is an exercise which is similar in nature, but discrete and simpler.

For the template we use the integers $\mathbb{Z}$ as a poset. There are two ways to do this, upwards or downwards. Here we use the downward version.

$$\cdots \longrightarrow 2 \longrightarrow 1 \longrightarrow 0 \longrightarrow -1 \longrightarrow -2 \longrightarrow \cdots$$

This helps with some of the calculations. Notice that the limit will occur at the positive end of $\mathbb{Z}$, and the colimit will occur at the negative end. We look at a simple diagram in **Top**, the category of topological spaces. We put the same space at each node, and the same map at each edge.

Let $\mathbb{O}$ be the circle group. We think of $\mathbb{O}$ as the circle in the cartesian plane with radius 1 and centre at the origin. Each point on $\mathbb{O}$ is determined by its unique polar coordinate θ with $0 \leq \alpha < 2\pi$. Here addition (mod 2π) is important. Put a copy of $\mathbb{O}$ at each index. For each edge

$$\mathbb{O} \xrightarrow{\quad \delta \quad} \mathbb{O}$$

we take the doubling map

$$\delta(\alpha) = 2\alpha \quad (\text{mod } 2\pi)$$

for co-ordinate α. This function wraps the source circle twice round the target circle. What can a left solution be?

The limit

The limit is some kind of topological space A furnished with a family of functions

$$A \xrightarrow{\quad \phi_m \quad} \mathbb{O} \qquad m \in \mathbb{N}$$

such that

$$\delta \circ \phi_{m+1} = \phi_m$$

for each $m \in \mathbb{N}$. Thus, for each $a \in A$ we have

$$\phi_m(a) = 2\phi_{m+1}(a) \quad (\text{mod } 2\pi)$$

and hence

$$\phi_m(a) = 2^r \phi_{m+r}(a) \quad (\text{mod } 2\pi)$$

for each $m, r \in \mathbb{N}$. Since we can divide any real number by 2, we seem to have

$$\phi_r(a) = \frac{1}{2^r} \phi_0(a)$$

for all $a \in A, r \in \mathbb{N}$. This is not right. We have forgotten the 2π aspect.

Suppose we have

$$\phi_0(a) = \alpha$$

for some $0 \le \alpha < 2\pi$. Then one of

$$\phi_1(a) = \frac{\alpha}{2} \qquad \phi_1(a) = \frac{\alpha}{2} + \pi = \frac{\alpha + 2\pi}{2}$$

must hold, and these can arise from

$$\phi_2(a) = \frac{\alpha}{4} \quad \phi_2(a) = \frac{\alpha + 4\pi}{4} \qquad \phi_2(a) = \frac{\alpha + 2\pi}{4} \quad \phi_2(a) = \frac{\alpha + 6\pi}{4}$$

and so on. These possibilities are conveniently displayed as a tree

$$
\begin{array}{cccccccc}
\vdots & \vdots & \vdots & \vdots & \vdots & \vdots & \vdots & \vdots \\
\dfrac{\alpha}{8} & \dfrac{\alpha+8\pi}{8} & \dfrac{\alpha+4\pi}{8} & \dfrac{\alpha+12\pi}{8} & \dfrac{\alpha+2\pi}{8} & \dfrac{\alpha+10\pi}{8} & \dfrac{\alpha+6\pi}{8} & \dfrac{\alpha+14\pi}{8}
\end{array}
$$

$$
\dfrac{\alpha}{4} \qquad\qquad \dfrac{\alpha+4\pi}{4} \qquad\qquad \dfrac{\alpha+2\pi}{4} \qquad\qquad \dfrac{\alpha+6\pi}{4}
$$

$$
\dfrac{\alpha}{2} \qquad\qquad\qquad\qquad \dfrac{\alpha+2\pi}{2}
$$

$$
\alpha
$$

and each $a \in A$ with $\phi_0(a) = \alpha$ generates a branch of the tree. Each point of A corresponds to a certain region of $\mathbb{O}$.

The colimit

The colimit is easier, and we can give a full description of it.

We first look at some properties of a general right solution.

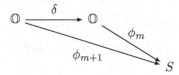

Thus we consider a space S furnished with a $\mathbb{Z}$ indexed family of functions ϕ_m such that each indicated triangle commutes. In other words

$$2\alpha \equiv \beta \implies \phi_{m+1}(\alpha) = \phi_m(\beta)$$

for all coordinates α, β and all $m \in \mathbb{Z}$. Here and below

$$\lambda \equiv \rho \quad \text{means} \quad \lambda = \rho \pmod{2\pi}$$

with all the usual properties. By a trivial induction the equivalence gives

$$2^r \alpha \equiv \beta \Longrightarrow \phi_{k+r}(\alpha) = \phi_k(\beta)$$

for all coordinates α, β, all $k \in \mathbb{Z}$, and all $r \in \mathbb{N}$. We need a variant of this last result, namely

$$(\star) \quad 2^m \alpha \equiv \beta \Longrightarrow \phi_m(\alpha) = \phi_0(\beta)$$

for all coordinates α, β and all $m \in \mathbb{Z}$. Notice how this shows that the map ϕ_0 determines the whole of the structure of the solution.

We now show that *any* function

$$\mathbb{O} \xrightarrow{\phi} S$$

to a space can be used to generate a solution. To prove this implication suppose that

$$2^m \alpha \equiv \beta$$

holds. If $m \le 0$ then, with $r = m$ the previous result gives

$$\phi_{k+m}(\alpha) = \phi_k(\beta)$$

for all $k \in \mathbb{Z}$. Thus we take $k = 0$. Suppose $m \le 0$, say $m = -r$ for $r \in \mathbb{N}$. Then

$$0 \le 2^m \alpha = 2^{-r} \alpha \le \alpha$$

so that

$$2^{-r} \alpha = \beta$$

and hence

$$\alpha = 2^r \beta$$

holds. A version of the previous result now gives

$$\phi_{k+r}(\beta) = \phi_k(\alpha)$$

for all $k \in \mathbb{Z}$. Thus taking $k = m$ gives the required result.

Consider the maps

$$\mathbb{O} \xrightarrow{\rho_m} \mathbb{O}$$

given by

$$\rho_m(\alpha) \equiv 2^m \alpha$$

for all $m \in \mathbb{Z}$ and coordinates α. In particular, note that ρ_0 is the identity function on $\mathbb{O}$. We check that these maps furnish $\mathbb{O}$ as a right solution.

We have

$$\rho_{m+1}(\alpha) \equiv 2^{m+1}\alpha$$

for all $m \in \mathbb{Z}$ and coordinates α. Let

$$\beta \equiv 2\alpha$$

so that

$$2^m\beta \equiv 2^{m+1}\alpha$$

and hence

$$(\rho_m \circ \delta)(\alpha) = \rho_m(\beta) \equiv 2^m\beta \equiv \rho_{m+1}(\alpha)$$

so that

$$\rho_m \circ \delta = \rho_{m+1}$$

as required.

Finally, we show that $\mathbb{O}$ with these furnishings is the colimit of the diagram. To do that consider an arbitrary right solution, as above. We require a unique map

$$\mathbb{O} \xrightarrow{\quad\mu\quad} S$$

such that

$$\phi_m = \mu \circ \rho_m$$

for each $m \in \mathbb{Z}$. By considering the case $m = 0$ we see that $\mu = \phi_0$ is the only possible map. We check that this does mediate.

Consider any $m \in \mathbb{Z}$ and any coordinate α. With

$$\beta \equiv 2^m\alpha$$

we have

$$(\mu \circ \rho_m)(\alpha) = \mu(\beta) = \phi_0(\beta)$$

so that

$$\phi_m(\alpha) = \phi_0(\beta)$$

is the requirement. This is precisely the result $(\star)$.

After reading this you might think this colimit example is a bit of a cheat. You could be right.

Exercises

4.5.1 For the template use $\mathbb{Z}$ as a poset. It doesn't matter which way you order $\mathbb{Z}$, but you might find it easier to use the upwards version, positives to the right and negatives to the left.

Consider the following diagram in *Set*. At each node place $\mathbb{Z}$ (so now $\mathbb{Z}$ is playing two different roles). At each edge place the doubling function.

$$\mathbb{Z} \xrightarrow{\ d\ } \mathbb{Z}$$
$$z \longmapsto 2z$$

Show that the arrows of any left solution have a simple behaviour.

Describe the limit of the diagram. You should verify your claims.

Investigate the behaviour of an arbitrary right solution.

Show that when suitably furnished, the dyadic rationals are the colimit of the diagram.

4.5.2 For the template use $\mathbb{Z}$ as a linear order, and for convenience let the edges point to the right, the positive end of $\mathbb{Z}$. We look at a $\mathbb{Z}$-diagram in *Pos*, the category of posets.

At each node i place a copy of the 3-element poset $A(i)$

with a top, bottom, and a middle elements $\star$. For each successive edge $(i+1, i)$ of the template $\mathbb{Z}$, from node i to node $i + 1$, consider the monotone map

that sends the three source elements to the central target element. This gives the $\mathbb{Z}$ diagram in *Pos*.

Describe both the limit and the colimit of this diagram.

4.6 How to calculate certain limits

For many categories the objects are structured sets, and the arrows are structure preserving functions. Each object is a single set, its carrier, furnished with some structure, perhaps restricted by certain required properties (axioms). Each arrow is a function between the carriers which respects the structure in an appropriate fashion. For instance

$$Pos \qquad Mon \qquad Top$$

are three categories of this kind. For each such category C there is a forgetful functor

$$C \longrightarrow Set$$

which sends each object to its carrier, and views each arrow as the function. Often this functor is named U, for underlying.

In this section we show how to compute a limit in such a category C. To do that we first compute the corresponding limit in Set, and then lift back up to C. Thus we must first show how to calculate limits in Set. The construction we give works for many categories of structured sets, but not necessarily all. In the next section we look at certain colimits.

We recall the notation we have used throughout this chapter. We assume we have a diagram in the category under investigation given by a template ∇. We let $i, j, k, \ldots$ range over the family $\mathbb{I}$ of nodes of ∇, and we let $e, f, g, \ldots$ range over the family $\mathbb{E}$ of edges of ∇. The construction works when ∇ is a graph and when ∇ is a category.

4.6.1 Limits in Set

In this block we show how to compute the limit of a diagram in Set. This then forms the basis for limits in the other three and similar categories.

We begin with a review of products in Set. Binary products are easy, we take the cartesian product – the set of ordered pairs – of the two component sets. Products of finitely many components are just as easy, again we take the cartesian product of the components.

What about the product of an arbitrary indexed family

$$\mathsf{A} = \big(A(i) \,|\, i \in \mathbb{I} \big)$$

of sets? Let

$$\bigcup \mathsf{A}$$

be the union of the family A. (It is often useful to tag each component $A(i)$

so we have a disjoint union. But that is not needed here.) We look at certain functions

$$\mathbb{I} \longrightarrow \bigcup A$$

from nodes to elements.

4.6.1 Definition Let A be an $\mathbb{I}$-indexed family of sets, as above. A choice function for A is a function

$$a(\cdot) : \mathbb{I} \longrightarrow \bigcup A \quad \text{such that} \quad a(i) \in A(i)$$

for each node $i \in \mathbb{I}$. $\qquad\qquad\square$

A choice function selects one member from each component set $A(i)$. When $\mathbb{I}$ is finite such a choice function can be coded as a tuple. When $\mathbb{I}$ is not finite we need this more general idea.

4.6.2 Definition Let A be an $\mathbb{I}$-indexed family of sets, let $\prod A$ be the set of all choice functions for A, and for each node $i \in \mathbb{I}$ let

$$\prod A \xrightarrow{\;\alpha(i)\;} A(i)$$
$$a \longmapsto a(i) \qquad \alpha(i)(a) = a(i)$$

be the 'evaluation at i' function. $\qquad\qquad\square$

Before you continue reading you might try to show that

$$(\prod A \xrightarrow{\;\alpha(i)\;} A(i) \mid i \in \mathbb{I})$$

is a product wedge in *Set*. Here we prove something more general.

We have a template ∇ with nodes $\mathbb{I}$ and edges $\mathbb{E}$. Suppose also we have a diagram in *Set*

$$\mathsf{A} = \big(A(i) \mid i \in \mathbb{I}\big) \qquad \mathcal{A} = \big(A(e) \mid e \in \mathbb{E}\big)$$

which instantiates the template. We produce a limit of this diagram.

4.6.3 Definition Given a ∇-diagram $(\mathsf{A}, \mathcal{A})$, as above, a thread is a choice function

$$a(\cdot) : \mathbb{I} \longrightarrow \bigcup A$$

such that

$$A(e)\big(a(i)\big) = a(j) \quad \text{for each edge} \quad i \xrightarrow{\;e\;} j$$

of ∇. $\qquad\qquad\square$

A choice function merely selects a member from each component set $A(i)$. A thread ensures that these selections are compatible. If we pass from one component $A(i)$ to another $A(j)$ using an edge $A(e)$, then we can take the selected element with us knowing that we will arrive at the selected element at the end.

If $\mathbb{E} = \emptyset$ then every choice function is a thread.

4.6.4 Definition   Given a ∇-diagram $(\mathsf{A}, \mathcal{A})$, as above, let A be the set of all threads, and for each node $i \in \mathbb{I}$ let

$$A \xrightarrow{\;\alpha(i)\;} A(i)$$
$$a \longmapsto a(i) \qquad \alpha(i)(a) = a(i)$$

be the 'evaluation at i' function. $\square$

By the observation before the definition, this does not conflict with Definition 4.6.2. It extends the idea to a more general context.

4.6.5 Lemma   *Given a ∇-diagram $(\mathsf{A}, \mathcal{A})$, as above, the evaluation functions furnish the set of threads as a left solution of the diagram.*

Proof   We must show that for each edge e the induced triangle,

$$\begin{array}{ccc} & \alpha(i) \nearrow\; A(i) & i \\ A & \Big\downarrow A(e) & e \\ & \alpha(j) \searrow\; A(j) & j \end{array} \qquad \big(A(e) \circ \alpha(i)\big)(a) = \alpha(j)(a)$$

as on the left, commutes. In other words we require the equality on the right for each $a \in A$. But since a is a thread we have

$$\big(A(e) \circ \alpha(i)\big)(a) = \big(A(e)\big(\alpha(i)(a)\big) = \big(A(e)\big(a(i)\big) = a(j) = \alpha(j)(a)$$

as required. $\square$

With this we have the result we want.

4.6.6 Theorem   *Given a ∇-diagram $(\mathsf{A}, \mathcal{A})$, as above, the evaluation functions furnish the set of threads as a limit of the diagram.*

Proof   By Lemma 4.6.5 we already know that we have a left solution of the diagram. Thus it suffices to show that this left solution is universal.

To this end let

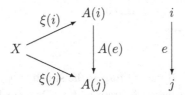

be a typical part of an arbitrary left solution of the diagram. We require a unique function

$$X \xrightarrow{\mu} A \quad \text{such that} \quad \xi(i) = \alpha(i) \circ \mu$$

for each $i \in \mathbb{I}$.

To obtain one such function set

$$\mu(x)(i) = \xi(i)(x)$$

for each $x \in X$ and $i \in \mathbb{I}$. Since

$$\xi(i)(x) \in A(i)$$

we see that $\mu(x)$ is a choice function. For each edge e, as above, we have

$$A(e)\big(\mu(x)(i)\big) = A(e)\big(\xi(i)(x)\big) = \big(A(e) \circ \xi(i)\big)(x) = \xi(j)(x) = \mu(x)(j)$$

to show that $\mu(x)$ is a thread. Thus we do have a function μ of the required type. Finally, for each node i we have

$$\big(\alpha(i) \circ \mu\big)(x) = \alpha(i)\big(\mu(x)\big) = \mu(x)(i) = \xi(i)(x)$$

to show that

$$\alpha(i) \circ \mu = \xi(i)$$

as required.

This deals with the existence of a mediating arrow. Now we must deal with the uniqueness. To this end suppose we have a function ν with

$$\alpha(i) \circ \nu = \xi(i)$$

for each node i. Consider any $x \in X$ and the corresponding thread $a = \nu(x)$. For each node i we have

$$\begin{aligned}
\nu(x)(i) &= a(i) \\
&= \alpha(i)(a) \\
&= \alpha(i)\big(\nu(x)\big) \\
&= \big(\alpha(i) \circ \nu\big)(x) \quad = \xi(i)(x) = \mu(x)(i)
\end{aligned}$$

to show $\nu = \mu$, as required. $\qquad\square$

In the next three blocks we show how to calculate limits in certain categories C of structured sets. The process is the same for these categories. In each case there is a forgetful functor

$$C \longrightarrow Set$$

to the category of sets. It merely forgets the structure. Given a diagram in C this functor converts it into a diagram in Set. We calculate the limit of that Set-diagram using the method of this block. The problem then is to furnish that set and collection of arrows so that they form a limit in C. This last part needs some special properties of C.

Exercises

4.6.1 Let $SetD$ be the category of sets with a distinguished subset, as used in Exercise 1.2.2. Thus each object is a pair (A, R) where A is a set and R is a subset $R \subseteq A$. The arrows are those functions which preserve the selected subset.

Let $\nabla = (\mathbb{I}, \mathbb{E})$ be a template, and consider a ∇-diagram

$$\mathsf{A}(D) = \big((A(i), R(i)) \,|\, i \in \mathbb{I}\big) \qquad \mathcal{A}(D) = \big(A(e) \,|\, e \in \mathbb{E}\big)$$

in $SetD$. By forgetting the distinguished subsets we have a ∇-diagram

$$\mathsf{A} = \big(A(i) \,|\, i \in \mathbb{I}\big) \qquad \mathcal{A} = \big(A(e) \,|\, e \in \mathbb{E}\big)$$

in Set. Consider the limit of the diagram in Set, that is the set of threads with the attached functions. Show that this can be furnished to produce a limit of the diagram in $SetD$.

4.6.2 Limits in Pos

We look at the category Pos of posets and monotone maps. We continue with the notation of the previous block. Thus we have a template ∇ with a collection $\mathbb{I}$ of nodes and a collection $\mathbb{E}$ of edges. These index certain objects and arrows in Pos. We assume we have an instantiation of ∇

$$\mathsf{A} = \big(A(i) \,|\, i \in \mathbb{I}\big) \qquad \mathcal{A} = \big(A(e) \,|\, e \in \mathbb{E}\big)$$

to form a diagram in Pos. Thus each $A(i)$ is a poset, and for each edge

$$i \xrightarrow{\ e\ } j$$

the arrow

$$A(i) \xrightarrow{\ A(e)\ } A(j)$$

is a monotone map. Eventually we put these extra facilities to good use.

For the first step we forget the extra facilities and drop down to **Set**. We look at the limit of this **Set**-diagram, the *set* of all threads

$$a : \mathbb{I} \longrightarrow \bigcup A$$

together with the $\mathbb{I}$-indexed family

$$A \xrightarrow{\ \sigma(i)\ } A(i)$$
$$a \longmapsto a(i)$$

of evaluation functions. Recall that a thread satisfies

$$A(e)\big(a(i)\big) = a(j)$$

for each edge e, as above. This family of equalities rephrases as

$$A(e) \circ \alpha(i) = \alpha(j)$$

using the evaluation function α.

Our job is to furnish A as a poset, and check that each evaluation function is monotone. In this way we produce a left solution of the **Pos**-diagram. We then show that this solution is universal, a limit in **Pos**.

How can we partially order A? In other words, how can we compare threads? In this kind of situation there is one trick that should always be tried. We use the pointwise comparison. We let

$$a \le b \Longleftrightarrow (\forall i \in \mathbb{I})[a(i) \le b(i)]$$

for threads a and b. Notice how this works. For threads a and b we pass to each poset $A(i)$ in turn and carry out a comparison there. All of these must give a positive answer.

Why does this give a partial order on A? Verifying the three properties is routine. Let's look at the antisymmetry. Consider two threads a and b with $a \le b \le a$. Then, for each $i \in \mathbb{I}$, we have

$$a(i) \le b(i) \le a(i)$$

and hence $a(i) = b(i)$, since $A(i)$ is a poset. Thus $a = b$.

Next we show that each evaluation function $\alpha(i)$ is monotone, that is

$$a \le b \Longrightarrow \alpha(i)(a) \le \alpha(i)(b)$$

for threads a and b. Since

$$\alpha(i)(a) = a(i) \qquad \alpha(i)(b) = b(i)$$

this is immediate.

This produces the furnishings. Why does it give us a left solution of the diagram in *Pos*? We require each cell

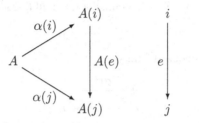

to be a commuting triangle in *Pos*, for each edge e, as indicated. This is certainly a triangle in *Pos*. We are given that each $A(e)$ is monotone, and we have ensured that $\alpha(i)$ and $\alpha(j)$ are monotone. This *Pos*-triangle commutes because it commutes down in *Set*.

Our next job is to show that this left solution is universal in *Pos*. To do that we compare it with an arbitrary left solution of the *Pos*-diagram. Thus we assume given a poset X together with an $\mathbb{I}$-indexed family of monotone maps

$$X \xrightarrow{\;\xi(i)\;} A(i)$$

such that the *Pos*-triangle

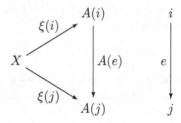

commutes for each edge e, as indicated. We require a unique mediator

$$X \xrightarrow{\;\mu\;} A$$

which, of course, must be monotone.

Think about this. If there is such a mediator μ then, by passing down to *Set*, it can only be that function that works for the *Set*-diagram. If that function turns out to be monotone, it will certainly make all the required *Pos*-triangles commute, for they commute down in *Set*. Thus we don't have much choice. The *Set*-mediator is given by

$$\mu(x)(i) = \xi(i)(x)$$

for each $x \in X$ and $i \in \mathbb{I}$. We have to show this function is monotone.

Why does

$$x \le y \implies \mu(x) \le \mu(y)$$

hold for all $x, y \in X$? Consider such $x, y \in X$ with $x \le y$. We are given that each $\xi(i)$ is monotone, so that

$$\xi(i)(x) \le \xi(i)(y)$$

and hence, by definition of μ, we have

$$\mu(x)(i) \le \mu(y)(i)$$

for each $i \in \mathbb{I}$. Finally, remember that each of $\mu(x)$ and $\mu(y)$ is a thread, so that this last universally quantified comparison gives

$$\mu(x) \le \mu(y)$$

as required.

The construction of this block is fairly typical. We will use it again with minor variations to produce limits in two more categories.

Exercises

4.6.2   (a) Let *Eqv* be the category of equivalence relations. Each object is a pair $(A, \sim)$ where A is a set and $\sim$ is an equivalence relation on A. Make sure you understand the arrows. Exercise 1.2.3 will help.

(b) Let $\nabla = (\mathbb{I}, \mathbb{E})$ be a template, and consider a ∇-diagram

$$\mathsf{A}(\sim) = \big((A(i), \sim_i) \,|\, i \in \mathbb{I} \big) \qquad \mathcal{A}(D) = \big(A(e) \,|\, e \in \mathbb{E} \big)$$

in *Eqv*. By forgetting the distinguished subsets we obtain a ∇-diagram

$$\mathsf{A} = \big(A(i) \,|\, i \in \mathbb{I} \big) \qquad \mathcal{A} = \big(A(e) \,|\, e \in \mathbb{E} \big)$$

in *Set*. Consider the limit of the diagram in *Set*, that is the set of threads with the attached functions. Show that this can be furnished to produce a limit of the diagram in *Eqv*.

4.6.3   Let $\nabla = (\mathbb{I}, \mathbb{E})$ be a template. Let R be a monoid and let

$$\mathsf{A} = \big(A(i) \,|\, i \in \mathbb{I} \big) \qquad \mathcal{A} = \big(A(e) \,|\, e \in \mathbb{E} \big)$$

be a ∇-diagram in *Set-R*. Show how the set of threads can be furnished to produce a limit of this diagram.

4.6.3 Limits in *Mon*

In this block we show how the construction of Block 4.6.1 also produces limits in *Mon*, the category of monoids. The general procedure is the same as outlined at the beginning of Block 4.6.2. Starting from a diagram in *Mon*, we pass to *Set* and take the limit of that *Set*-diagram. The main problem is to furnish that *Set*-limit to become a limit of the original *Mon*-diagram.

Recall that a monoid is a furnished set

$$(A, \bullet, 1)$$

where '$\bullet$' is a binary operation on A, and 1 is a distinguished element. These attributes must satisfy

$$(a \bullet b) \bullet c = a \bullet (b \bullet c) \qquad 1 \bullet a = a = a \bullet 1$$

for all $a, b, c \in A$. In general, we write

$$ab \quad \text{for} \quad a \bullet b$$

but there is one case when we will explicitly show the operation symbol.

Monoids are the objects of *Mon*. The arrows are the monoid morphism. Recall that a monoid morphism

$$(A, \bullet, 1) \xrightarrow{f} (B, \bullet, 1)$$

is a function

$$f : A \longrightarrow B$$

between the carriers such that

$$f(ab) = f(a)f(b) \qquad f(1) = 1$$

for all $a, b \in A$.

As usual we have a template ∇ with a collection $\mathbb{I}$ of nodes and a collection $\mathbb{E}$ of edges. We also have an instantiation

$$\mathsf{A} = \big(A(i) \mid i \in \mathbb{I}\big) \qquad \mathcal{A} = \big(A(e) \mid e \in \mathbb{E}\big)$$

to form a *Mon*-diagram. Thus each $A(i)$ is a monoid and each $A(e)$ is a monoid morphism. By passing to *Set* we obtain the set A of all threads

$$a : \mathbb{I} \longrightarrow \bigcup \mathsf{A}$$

together with the $\mathbb{I}$-indexed family

$$A \xrightarrow{\alpha(i)} A(i)$$

of evaluation functions. Our first job is to furnish A as a monoid, and check that each evaluation function is a monoid morphism.

How can we combine a pair of threads a and b to form

$$a \star b$$

a third thread? This is the case where we will explicitly indicate the operation. Let

$$(a \star b)(i) = a(i)b(i)$$

for each $i \in \mathbb{I}$. Notice how we use the operation on $A(i)$. This construction certainly gives a function

$$a \star b : \mathbb{I} \longrightarrow \bigcup A$$

and it is a choice function since $a(i)b(i)$ lives in $A(i)$.

Why is $a \star b$ a thread? Consider an arbitrary edge e from i to j. We remember that $A(e)$ is a monoid morphism. Thus

$$A(e)\big((a \star b)(i)\big) = A(e)\big(a(i)b(i)\big)$$
$$= \Big(A(e)\big(a(i)\big)\Big)\Big(A(e)\big(b(i)\big)\Big) = a(j)b(j) = (a \star b)(j)$$

as required. Why is this operation associative? Because

$$((a \star b) \star c)(i) = \big((a \star b)(i)\big)c(i)$$
$$= \big(a(i)b(i)\big)c(i)$$
$$= a(i)\big(b(i)c(i)\big)$$
$$= a(i)\big((b \star c)(i)\big) \quad = \big(a \star (b \star c)\big)(i)$$

for all $a, b, c \in A$ and $i \in \mathbb{I}$. We need a distinguished element of A. We set $1(i) = 1_i$ for each $i \in \mathbb{I}$. Here 1_i is the distinguished element of $A(i)$. It is easy to check that this function 1 is a thread and, almost trivially, we have

$$1 \star a = a = a \star 1$$

for each thread a. Thus we have furnished A as a monoid.

Why is each evaluation function

$$A \xrightarrow{\ \alpha(i)\ } A(i)$$

a monoid morphism? Because

$$\alpha(i)\big(a \star b\big) = (a \star b)(i) = a(i)b(i) = \big(\alpha(i)(a)\big)\big(\alpha(i)(b)\big)$$

for each $a, b \in A$.

This sets up the furnishings. Why does it give a left solution of the **Mon**-diagram? To be a left solution we require that certain triangles commute in **Mon**. These triangles do commutes in **Set**, and they are triangles in **Mon**, so they commute in **Mon**.

Our main job is to show that this left solution is universal in **Mon**. To do that we compare it with an arbitrary left solution of the **Mon**-diagram. We assume given a monoid X together with an $\mathbb{I}$-indexed family of monoid morphisms

$$x \xrightarrow{\quad \xi(i) \quad} A(i)$$

such that the **Mon**-triangle

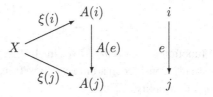

commutes for each edge e, as indicated. We require a unique mediator

$$X \xrightarrow{\quad \mu \quad} A$$

which, of course, must be a monoid morphism.

By passing to **Set** we see there is only one possible function μ we can use, that given by

$$\mu(x)(i) = \xi(i)(x)$$

for each $x \in X$ and $i \in \mathbb{I}$. Thus it suffices to show that this function μ is a monoid morphism. For each $x, y \in X$ and $i \in \mathbb{I}$, remembering that $\xi(i)$ is a monoid morphism, we have

$$\begin{aligned}
\mu(xy)(i) &= \xi(i)(xy) \\
&= \big(\xi(i)(x)\big)\big(\xi(i)(y)\big) \\
&= \big(\mu(x)(i)\big)\big(\mu(y)(i)\big) = \big(\mu(x) \star \mu(y)\big)(i)
\end{aligned}$$

so that

$$\mu(xy) = \mu(x) \star \mu(y)$$

to show that μ passes across the operation. The other requirement

$$\mu(1) = 1$$

is even easier.

This kind of construction works in many algebraic categories. In fact, the earlier Exercise 4.6.3 uses exactly the same technique.

Exercises

4.6.4 (a) Let **CMon** be the category of commutative monoids. Show that the construction of this block produces limits in **CMon**.

(b) Let **Grp** be the category of groups. Show that the construction of this block produces limits in **Grp**.

(c) Let **Rng** be the category of unital rings. Show that **Rng** has limits for $\nabla = (\mathbb{I}, \mathbb{E})$ diagrams.

4.6.5 A partially ordered monoid (a pom) is a structure

$$(A, \leq, \cdot, 1)$$

where $(A, \leq)$ is a poset and $(A, \cdot, 1)$ is a monoid, and

$$\left.\begin{array}{c} x \leq a \\ y \leq b \end{array}\right\} \implies xy \leq ab$$

for all $a, b, x, y \in A$. These are the objects of the category **Pom**. An arrow of **Pom** is a monoid morphism that is also monotone. For an arbitrary template $\nabla = (\mathbb{I}, \mathbb{E})$ show that each ∇-diagram in **Pom** has a limit.

4.6.4 Limits in *Top*

In this block we show how the construction of Block 4.6.1 also produces limits in *Top*, the category of topological spaces and continuous maps. In topological circles a left limit is usually called an inverse limit. More often than not the template is a partial order or even a pre-order, and the indexing is contravariant. We won't deal with that aspect here.

As usual we have a template ∇ of nodes $\mathbb{I}$ and edges

$$\mathsf{A} = \big(A(i) \,|\, i \in \mathbb{I}\big) \qquad \mathcal{A} = \big(A(e) \,|\, e \in \mathbb{E}\big)$$

to form a *Top*-diagram. Thus each $A(i)$ is a topological space and each $A(e)$ is a continuous map. By passing to *Set* we obtain the set A of all threads

$$a : \mathbb{I} \longrightarrow \bigcup \mathsf{A}$$

together with the $\mathbb{I}$-indexed family

$$A \xrightarrow{\ \alpha(i)\ } A(i)$$

of evaluation functions. Our first job is to convert A into a topological space in such a way that each $\alpha(i)$ is continuous.

Consider any node $i \in \mathbb{I}$ and any open $U \in \mathcal{O}A(i)$ of that component. We certainly require the inverse image set

$$i(U) = \alpha(i)^{\leftarrow}(U) = \{a \in A \,|\, a(i) \in U\}$$

to be open. Thus we take the family of all these subsets of A as a subbase of a topology on A. This ensures that each $\alpha(i)$ is continuous.

In the usual way this gives us a left solution of the diagram. This has nothing much to do with the topological aspects. It's merely that certain triangles of function do commute.

Our main job is to show that this left solution is universal in **Top**. To do that we compare it with an arbitrary left solution of the **Top**-diagram. Thus we assume given a topological space X together with an $\mathbb{I}$-indexed family of continuous maps

$$X \xrightarrow{\;\xi(i)\;} A(i)$$

such that the **Top**-triangle

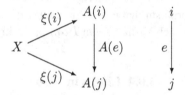

commutes for each edge e, as indicated. We require a unique mediator

$$X \xrightarrow{\;\mu\;} A$$

which, of course, must be a continuous map.

By passing to **Set** the only possible mediating function is given by

$$\mu(x)(i) = \xi(i)(x)$$

for each $x \in X$ and $i \in \mathbb{I}$. Thus it suffices to show that this function μ is continuous. To do that it suffices to show that for each subbasic open set of A the inverse image across μ is open. Thus we require

$$\mu^{\leftarrow}(i(U)) \in \mathcal{O}R$$

for each node $i \in \mathbb{I}$ and open $U \in \mathcal{O}A(i)$. For each $x \in X$ we have

$$x \in \mu^{\leftarrow}(i(U)) \Longleftrightarrow \mu(x) \in i(U) = \alpha(i)^{\leftarrow}(U)$$
$$\Longleftrightarrow \alpha(i)(\mu(x)) \in U$$
$$\Longleftrightarrow \mu(x)(i) \in U$$
$$\Longleftrightarrow \xi(i)(r) = \mu(x)(i) = U \quad \Longleftrightarrow x \in \xi(i)^{\leftarrow}(U)$$

to show that

$$\mu^{\leftarrow}(i(U)) = \xi(i)^{\leftarrow}(U)$$

for each pair i and U. Since each $\xi(i)$ is continuous this shows that each $\mu^{\leftarrow}(i(U))$ is open, for the required result.

Notice how this construction and proof works. To ensure that each $\alpha(i)$ is continuous we need at least all the sets $i(U)$ in the topology on A. To show that a mediator μ is continuous we can't deal with more than the sets $i(U)$.

Exercises

4.6.6 Let $\mathbb{I}$ be a discrete template (that is, just a set). Let A be an $\mathbb{I}$-diagram in **Top** (that is, an $\mathbb{I}$-indexed family of topological spaces). Describe the limit of A, and relate this to a standard topological notion.

4.7 Confluent colimits in *Set*

We have looked at examples of limits. In this section we see how to calculate a certain kind of colimit. We work in the category *Set* of sets, but the same method works for other categories of structured sets.

For this example we assume the template is a poset. Thus let $\mathbb{I}$ be a poset with nodes

$$i, j, k, \ldots$$

and for each comparison $i \leq j$ let

$$i \xrightarrow{\ (j, i)\ } j$$

be the corresponding edge.

We assume $\mathbb{I}$ satisfies a certain restriction

4.7.1 Definition The poset $\mathbb{I}$ is **directed** if for each $i, j \in \mathbb{I}$ there is some $k \in \mathbb{I}$ with $i, j \leq k$.

The poset $\mathbb{I}$ is **confluent** if for each $i, j, l \in \mathbb{I}$ with $l \leq i, j$, there is some $k \in \mathbb{I}$ with $i, j \leq k$. $\square$

Trivially, each directed poset is confluent, but there are confluent posets that are not directed. For example, each discrete set is confluent but not directed (if it has more than one node). Notice that if a poset is directed then (by repeated use of this property) each finite subset has at least one upper bound. Similarly, if a poset is confluent, then each finite subset which has a lower bound also has an upper bound.

We assume the template poset $\mathbb{I}$ is confluent.

Let $(\mathsf{A}, \mathcal{A})$ be an $\mathbb{I}$-diagram in *Set*. This is an $\mathbb{I}$-indexed family of sets

$$A(i)$$

together with connecting functions

$$A(i) \xrightarrow{A(j, i)} A(j)$$

one for each comparison $i \leq j$ in $\mathbb{I}$. These functions must compose in the usual way, that is

$$A(i, i) = id_{A(i)} \quad \text{and} \quad A(k, j) \circ A(j, i) = A(k, i)$$

for $i \leq j \leq k$. To help with the later calculations it is convenient to write

$$
\begin{aligned}
A(i) &\longrightarrow A(j) \\
x &\longmapsto x|j
\end{aligned}
$$

for the function $A(j, i)$. Thus

$$x|j = A(j, i)(x)$$

and we may think of this as the 'restriction' of $x \in A(i)$ to j. Note that

$$x|i = x \qquad x|j|k = x|k$$

for $i \leq j \leq k$ with $x \in A(i)$.

To obtain the colimit of $(\mathsf{A}, \mathcal{A})$ we first produce the coproduct of A. Consider the disjoint union of the sets $A(i)$. We set this up with some care. We let

$$\amalg \mathsf{A} = \bigcup \{ A(i) \times \{i\} \mid i \in \mathbb{I} \}$$

that is we take the set of all pairs

$$(x, i)$$

for $i \in \mathbb{I}$ and $x \in A(i)$. An element x may occur many times, but each occurrence is tagged with an index. For each $i \in \mathbb{I}$ there is a function

$$
\begin{aligned}
A(i) &\longrightarrow \amalg \mathsf{A} \\
x &\longmapsto (x, i)
\end{aligned}
$$

and these functions structure $\amalg \mathsf{A}$ as the coproduct of A in *Set*. The proof of that is the discrete case of the following proof.

For the general situation we take a quotient of IIA. Consider the relation $\sim$ on IIA given by

$$(x, i) \sim (y, j) \iff (\exists i, j \leq k)[x|k = y|k]$$

for (x, i) and (y, j) from IIA. Here the quantifier looks for a node k which is an upper bound of the two given nodes. Trivially, this relation is reflexive and symmetric. We show it is transitive, and hence is an equivalence relation on IIA. To do that we use the confluence of $\mathbb{I}$.

Consider any situation

$$(x, i) \sim (y, j) \sim (z, k)$$

so that $(x, i) \sim (z, k)$ is required. We are given nodes l, m with

$$i, j \leq l \qquad j, k \leq m$$
$$x|l = y|l \qquad y|m = z|m$$

respectively. Now look how the nodes sit in $\mathbb{I}$.

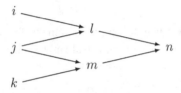

The nodes l, m have a lower bound and hence, by confluence, they have some upper bound n. But now

$$x|n = x|l|n = y|l|n = y|n = y|m|n = z|m|n = z|n$$

to show that

$$(x, i) \sim (z, k)$$

as required.

Since $\sim$ is an equivalence relation on IIA we may take

$$L = IIA/\sim$$

the family of blocks (equivalence classes) of $\sim$ in IIA. Let

$$IIA \longrightarrow L$$
$$(x, i) \longmapsto [x, i]$$

be the associated quotient function. For later observe that

$$[x, i] = [x|j, j]$$

for all nodes $i \leq j$.

We show that the family of composite functions

$$A(i) \longrightarrow \amalg A \longrightarrow L$$
$$x \longmapsto (x, i) \longmapsto [x, i]$$

structure L as the colimit of the $\mathbb{I}$-diagram A.

We first show that we do have a right solution of the diagram. Consider any pair of nodes $i \leq j$. We require that the inner triangle commutes.

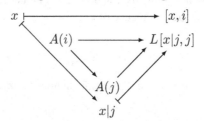

To prove that we track an element from the top left hand corner to the top right hand corner by the short trip and the long trip. We require

$$[x, i] = [x|j, j]$$

and this is nothing more than the observation made above.

Now we show that we have a universal right solution. Consider any right solution M where for nodes $i \leq j$ the commuting triangle

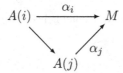

is a typical part of the structure. Thus we have

$$\alpha_j(x|j) = \alpha_i(x)$$

for each $x \in A(i)$. We require a unique function

$$L \xrightarrow{\ \mu\ } M$$

such that for each node i the triangle

$$A(i) \xrightarrow{\ \alpha_i\ } M$$

commutes. If there is such a function μ, then

$$\mu([x, i]) = \alpha_i(x)$$

for each $i \in \mathbb{I}$ and $x \in A(i)$. This shows that there is at most one such μ. To show there is at least one, it suffices to show that this assignment is well-defined, that is

$$[x, i] = [y, j] \implies \alpha_i(x) = \alpha_j(y)$$

for $i, j \in \mathbb{I}$ and $x \in A(i), y \in A(j)$. Assuming

$$[x, i] = [y, j]$$

we have

$$(x, i) \sim (y, j) \quad \text{to give} \quad x|k = y|k$$

for some node k with $i, j \leq k$. But now

$$\alpha_i(x) = \alpha_k(x|k) = \alpha_k(y|k) = \alpha_j(y)$$

for the required result.

Exercises

4.7.1 Let $\mathbb{I}$ be a confluent poset and let $(A, \mathcal{A})$ be an $\mathbb{I}$-diagram in *Set-R* for some monoid R. Thus for each node i the component $A(i)$ is an R-set, and for each pair $i \leq j$ of nodes the function

$$A(i) \xrightarrow{A(j, i)} A(j)$$

is an R-morphism. Show that A has a colimit in *Mon*.

4.7.2 Let $\mathbb{I}$ be a confluent poset and let $(A, \mathcal{A})$ be an $\mathbb{I}$-diagram in *Mon*. Thus for each node i the component $A(i)$ is a monoid, and for each pair $i \leq j$ of nodes the function

$$A(i) \xrightarrow{A(j, i)} A(j)$$

is a monoid morphism. Show that A has a colimit in *Mon*.

4.7.3 The solution to the last part of Exercise 2.5.2 is a bit terse. We can now fill in the missing details. Consider any category of algebras such as *Mon*. We produce the coproduct $A \amalg B$ of two objects A, B in three stages. First forget the carried structure and look at the coproduct in *Set*, the disjoint union $A \dot\cup B$. Next freely generate a monoid from this set. Finally take a quotient of this free monoid to convert certain functions into morphisms. Fill in the details.

4.7.4 Once you have mastered Exercise 4.7.3, show how to construct co-equalizers in *Pos*.

5

Adjunctions

The isolation of the notion of an adjunction is one of the most important contributions of category theory. In a sense adjoints form the first 'non-trivial' part of category theory; at least it can seem that way now that all the basic stuff has been sorted out. There are adjunctions all over mathematics, and examples were known before the categorical notion was formalized. We have already met several examples, and later I will point you to them.

In this chapter we go through the various aspects of adjunctions quite slowly. We look at each part in some detail but, I hope, not in so much detail that we lose the big picture.

There is a lot going on in adjunctions, and you will probably get confused more than once. You might get things mixed up, forget which way an arrow is supposed to go, not be able to spell contafurious, and so on. Don't worry. I've been at it for over 40 years and I still can't remember some of the details. In fact, I don't try to. You should get yourself to the position where you can recognize that perhaps there is an adjunction somewhere around, but you may not be quite sure where. You can then look up the details. If you ever have to use adjunctions every day, then the details will become second nature to you.

5.1 Adjunctions defined

When first seen in full categorical adjunctions can seem a bit daunting. There is a lot going on but in many particular examples much of this complexity can disappear. In this section we look at all the various components and eventually arrive at the formal definition. Then in the later sections we analyse the content of the notion. For most of this section we merely run through the various bits of gadgetry that make up an adjunction. We don't look at the restrictions imposed on these gadgets.

An adjunction is an interaction between two categories

$$\textbf{\textit{Src}} \qquad\qquad \textbf{\textit{Trg}}$$

which we may think of as the

source   target

category. This is merely a conventional distinction, but it can help. These categories play similar roles, but some of their attributes are in mirror image.

We have two covariant functors

$$\textbf{\textit{Src}} \underset{G}{\overset{F}{\rightleftarrows}} \textbf{\textit{Trg}}$$

going between the two categories, but in opposite directions. These functors are related in a certain way. We look at the details later. By convention we call

F the left G the right

adjoint of the pair, and we write

$$F \dashv G$$

to indicate this relationship. We also write any of

$$\textbf{\textit{Src}} \overset{F}{\underset{G}{\rightleftarrows}}{}^{\dashv} \textbf{\textit{Trg}} \qquad \textbf{\textit{Src}} \overset{F}{\underset{G}{\rightleftarrows}}{}^{\perp} \textbf{\textit{Trg}} \qquad \textbf{\textit{Src}} \overset{F \dashv G}{\rightleftarrows} \textbf{\textit{Trg}}$$

to indicate the relationship. Usually the left functor is placed above the right functor.

There are other notations used with adjunctions. Sometimes we write

$$F^* \quad \text{for} \quad F \qquad\qquad F_* \quad \text{for} \quad G$$

where the position of the decoration indicates which is the left and which is the right component. This can be useful when there is more than one adjunction around. In these circumstances we sometimes write F for the pair (F^*, F_*). For the time being we will stick with $F \dashv G$, but eventually we will use some of this other notation.

By convention we think of an adjunction as passing in the direction of its left adjoint. We write any of

$$\textbf{\textit{Src}} \longrightarrow \textbf{\textit{Trg}} \qquad \textbf{\textit{Src}} \overset{F \dashv G}{\longrightarrow} \textbf{\textit{Trg}} \qquad \textbf{\textit{Src}} \overset{F^* \dashv F_*}{\longrightarrow} \textbf{\textit{Trg}}$$

for the adjunction making use of the harpoon arrow to alert us. This is the conventional terminology and notation. Some of the older literature was written before these conventions were established. You might find that left and right are called something else, such as right and left.

The two functors can be composed to give endo-functors

$$G \circ F \text{ on } \textbf{\textit{Src}} \qquad\qquad F \circ G \text{ on } \textbf{\textit{Trg}}$$

respectively. These composites are related to the corresponding identity functors by natural transformations

$$Id_{\textbf{\textit{Src}}} \xrightarrow{\ \eta\ } G \circ F \qquad\qquad F \circ G \xrightarrow{\ \epsilon\ } Id_{\textbf{\textit{Trg}}}$$

called the

unit   counit

of the adjunction. Notice the left/right antisymmetry of these gadgets.

Consider any pair A and S of objects

$$A \text{ from } \textbf{\textit{Src}} \qquad\qquad S \text{ from } \textbf{\textit{Trg}}$$

respectively. An adjunction tries to compare these objects. Of course, there isn't a direct comparison, for they live in different worlds. To compare them we move one of the objects to the other category, and do the comparison there. Thus we use one of the two arrow sets

$$\textbf{\textit{Src}}[A, GS] \qquad\qquad \textbf{\textit{Trg}}[FA, S]$$

of the indicated category. Notice that the

left functor F right functor G

occurs in the

left position   right position

of the appropriate arrow set. This helps us to remember which functor is doing which job.

Each of the two arrow sets provides a place where we might compare the two objects. But which place should we use? It doesn't matter, for part of the gadgetry of an adjunction is an inverse pair of bijections

$$
\begin{array}{ccc}
f \longmapsto & \longrightarrow & f^{\sharp} \\
\textbf{\textit{Src}}[A, GS] & & \textbf{\textit{Trg}}[FA, S] \\
g_{\flat} & \longleftarrow & \longmapsto g
\end{array}
$$

between the two arrow sets. Furthermore, and this is the crucial aspect of an adjunction, these two assignments

$$(\cdot)^\sharp \qquad\qquad (\cdot)_\flat$$

must be natural for variation of both A and S.

There is a lot going on here. In the next section we work through this data again to see exactly what it means. Here is the formal definition.

5.1.1 Definition An adjunction

$$\left(F,\ G,\ (\cdot)^\sharp,\ (\cdot)_\flat\right)$$

consists of a pair of covariant functors

$$\textit{Src} \xrightarrow[\;\;G\;\;]{\;\;F\;\;} \textit{Trg}$$

where, for each

$$\textit{Src}\text{-object } A \qquad\qquad \textit{Trg}\text{-object } S$$

the two transposition assignments

$$\textit{Src}[A, GS] \xrightarrow[(\cdot)_\flat]{(\cdot)^\sharp} \textit{Trg}[FA, S]$$

form an inverse pair of bijections, and each is natural in A and S. □

Note that the data for an adjunction is not just a pair (F, G) of functors. We also need the transposition assignments $(\cdot)^\sharp$ and $(\cdot)_\flat$. Of course, we only need one of $(\cdot)^\sharp$ and $(\cdot)_\flat$, for each is the inverse of the other. The unit and the counit aren't mentioned in this definition, because we can show that

$$\eta_A = (id_{FA})_\flat \qquad\qquad \epsilon_S = (id_{GS})^\sharp$$

do those jobs. It turns out that various selections of the data

$$F \quad G \quad (\cdot)^\sharp \quad (\cdot)_\flat \quad \eta \quad \epsilon$$

can be put together in different ways to form an adjunction. We look at these various combinations in the following sections.

Exercises 5.1.1, 5.1.2, and 5.1.3 are rather straight forward. You have seen most of the gadgetry before. Exercises 5.1.4, 5.1.5, and 5.1.6 are more complicated, and you may have to return to them more than once.

Exercises

5.1.1   Each poset can be viewed as a category. Show that a poset adjunction, as in Example 1.3.3, is a categorical adjunction.

5.1.2   Each set S can be converted into a preset in two extreme ways. The discrete version uses equality as the comparison. For the indiscrete version any two elements are comparable. Show that the forgetful functor

$$Set \longleftarrow Pre$$

has both a left and a right adjoint, and these are different.

5.1.3   Show that the forgetful functor

$$Set \longleftarrow Top$$

has both a left and a right adjoint, and these are different.

5.1.4   For a poset S let $\mathcal{L}S$ be the poset of lower sections of S (under inclusion).

   (a) Show that each monotone function

$$S \xrightarrow{\ f\ } T$$

between posets induces a monotone function

$$\mathcal{L}S \xleftarrow{\ f^{\leftarrow}\ } \mathcal{L}T$$

via inverse image.

   (b) Show that the monotone function $f^{\leftarrow}$ has both a left adjoint and a right adjoint, and in general these are different. One of these is given by direct image, and the other isn't.

5.1.5   Let ∇ be a category viewed as a template, let C be an arbitrary category, and consider the diagonal functor.

$$C^{\nabla} \xleftarrow{\ \Delta\ } C$$

Show that C has a limit for each ∇-diagram precisely when Δ has a right adjoint, and sort out the corresponding result for colimits.

5.1.6   Let

$$Src \xleftarrow[G]{} Trg$$

be a functor which does have a left adjoint. Show that each limit cone in *Trg* is transported by G into a limit cone in *Src*.

5.1.7 Let

$$Src \xrightarrow{\quad F \dashv G \quad} Trg$$

be an adjunction, consider the product category $Src \times Trg$, and consider the two object assignments

$$(A, S) \xmapsto{\quad \mathfrak{F} \quad} Trg[FA, S]$$
$$Src \times Trg \longrightarrow Set$$
$$(A, S) \xmapsto[\mathfrak{G}]{\quad\quad} Trg[A, GS]$$

to the category of sets. In what sense are these the object assignments of a pair of functors? Describe the arrow assignments.

5.2 Adjunctions illustrated

Definition 5.1.1 says what an adjunction is, but doesn't tell us very much. There are several hidden consequences of the definition. In the following sections we take the definition apart, and look at the various components of an adjunction in some detail. To do that it will help if there are some examples we can look at. There are some simple examples. Exercises 5.1.2 and 5.1.3 give some of these, and we will see a few more. In this section we look at two (or perhaps it's three) examples with a bit more content. The idea is that as you read the following sections you can use these examples to illustrate what is going on. Thus you shouldn't expect to understand these examples immediately. Keep coming back to them as you are learning the various aspects of adjunctions.

The two examples, an algebraic example and a topological example, are miniature versions of more involved, and quite important, adjunctions. I will say what these larger versions are, but at this stage we don't go into any details. There is also a rather simple set-theoretic example (as a preliminary for the topological example). This is, perhaps, the best example to start with. However, you should be careful with it. In some ways it is too simple to bring out all of the different properties that an adjunction can have.

You should remember that an adjunction has many different components, but in any particular example not all of these are important, and some can be trivial. When learning about adjunctions in general, or about a particular adjunction, it is a good idea to sort out all its little bits and pieces.

5.2.1 An algebraic example

An involution algebra is a structure

$$(A, (\cdot)^\bullet)$$

carried by a set A where $(\cdot)^\bullet$ is an involution on this carrier. In other words $(\cdot)^\bullet$ is a 1-placed operation on A with

$$a^{\bullet\bullet} = a$$

for each $a \in A$. An involution morphism between involution algebras

$$A \xrightarrow{\ \phi\ } B$$

is a function, as indicated, such that

$$\phi(a^\bullet) = \phi(a)^\bullet$$

for each $a \in A$. This gives us the category

$$\textbf{\textit{Inv}}$$

of involution algebras and involution morphisms.

Let

$$\textbf{\textit{Set}} \xleftarrow{\ U\ } \textbf{\textit{Inv}}$$

be the underlying functor, the forgetful functor that loses the involution. We show that U has adjoints on both sides

$$\textbf{\textit{Set}} \xleftarrow{\ U\ } \textbf{\textit{Inv}}$$

with Σ above and Π below

and these are quite different.

Before we begin to construct these adjoints let's see where this example comes from. There are three levels of generalization each of which is worth analysing in its own right.

For the first level recall the notion of a (right) R-set for a monoid R. An involution algebra is nothing more than an R-set for an appropriate monoid R. (Can you see which one?) The forgetful functor

$$\textbf{\textit{Set}} \longleftarrow \textbf{\textit{Set-}}R$$

has both a left adjoint and a right adjoint. We look at this in Chapter 6. For the next level consider an arbitrary morphism

$$S \xrightarrow{\ f\ } R$$

between monoids. This induces a functor

$$\textbf{\textit{Set-S}} \longleftarrow \textbf{\textit{Set-R}}$$

called restriction of scalars. This functor has both a left adjoint and a right adjoint. When S is the trivial monoid this example reduces to the previous one. Finally let f, as above, be a ring morphism. This induces a restriction of scalars functor between the module categories.

$$\textbf{\textit{Mod-S}} \longleftarrow \textbf{\textit{Mod-R}}$$

Again this has both a left adjoint and a right adjoint. In this case the left adjoint is given by a tensor product, and the right adjoint is an enriched hom-functor. All of these adjunctions are worth looking at some time.

Here we concentrate on the involutary case. From now on in this block the only algebras and morphisms we meet are involutary, so we drop the qualifier 'involution'.

To produce the two adjoints it will help if we let

Set		*Inv*
$X, Y, \ldots$	Objects	$A, B, \ldots$
$f, g, \ldots, k, \ldots$	Arrows	$\phi, \psi, \ldots, \lambda, \ldots$

range over the indicated gadgets.

We describe the initial parts of the two constructions in parallel.

For each set X let

$$\Sigma X = X + X \qquad \Pi X = X \times X$$

that is

$$\Sigma X = \{(x, i) \mid x \in X, \, i = 0, 1\} \qquad \Pi X = \{(x, y) \mid x, y \in X\}$$

the set of $\{0, 1\}$-tagged elements of X, and the set of ordered pairs from X. It is not hard to furnish each of these with an involution. For ΣX we flip the tag, and for ΠX we swap the components.

We require both Σ and Π to be functors to *Inv*. Given a function

$$Y \xrightarrow{\quad k \quad} X$$

what should

$$\Sigma Y \xrightarrow{\quad \Sigma(k) \quad} \Sigma X \qquad \Pi Y \xrightarrow{\quad \Pi(k) \quad} \Pi X$$

be? We soon find the answer, but verifying that it does produce morphisms takes a bit of work. Think about this before you continue.

This gives us two of the four components of each of the two adjunctions. Next we require an inverse pair of assignments

$$f \longmapsto f^\sharp \qquad\qquad \phi \longmapsto \phi^\sharp$$
$$Set[X, UA] \quad Inv[\Sigma X, A] \qquad Inv[A, \Pi X] \quad Set[UA, X]$$
$$\psi_\flat \longleftarrow\!\!\dashv \psi \qquad\qquad g_\flat \longleftarrow\!\!\dashv g$$

for each set X and each algebra A. This needs some thought.

For the left hand bijection it helps if we set

$$a^{(i)} = \begin{cases} a^\bullet & \text{if } i = 1 \\ a & \text{if } i = 0 \end{cases}$$

for each $a \in A$ and tag i. In particular we have

$$a^{(i)\bullet} = a^{(1-i)} = a^{\bullet(i)} \qquad \phi(a^{(i)}) = \phi(a)^{(i)}$$

for each $a \in A$, tag i, and morphism ϕ. For the right hand bijection remember that ϕ is a morphism. You should now think about these constructions for a while, and do Exercises 5.2.1 and 5.2.2.

Of course, this doesn't quite prove that each of

$$\left(\Sigma, U, (\cdot)^\sharp, (\cdot)_\flat \right) \qquad \left(U, \pi, (\cdot)^\sharp, (\cdot)_\flat \right)$$

is an adjunction. We need to show that each $(\cdot)^\sharp$ and each $(\cdot)_\flat$ is natural. However, let's leave that until we have a better idea of what that means.

Exercises

5.2.1 Describe the action of Σ and Π on functions, and verify that each result is a morphism.

5.2.2 Set up the two inverse pairs of bijections $(\cdot)^\sharp$ and $(\cdot)_\flat$. At this stage don't worry about the required naturality.

5.2.2 A set-theoretic example

In this block we describe an easy example of an adjunction where the category *Set* is both the source and target. We have met most of the components before. We do this example here because in the next block we produce an enriched version with *Set* replaced by *Top*. That has more content, but much of the gadgetry is the same as in this *Set* example.

Let I be a fixed set. We know that

$$- \times I$$

is an endo-functor on *Set*. We show that this functor has a right adjoint and this is another functor that we already know. It is the hom-functor $Set[I, -]$. Thus, we attach to each set Y the set of all functions

$$I \longrightarrow Y$$

from I to Y. To do that we use what at first may seem an odd notation. For each set Y let

$$I \Rightarrow Y$$

be the set of all functions from I to Y. This gives us two endo-functors

$$- \times I \qquad I \Rightarrow -$$

on Set. Recall that each function

$$X_2 \xrightarrow{\ k\ } X_1 \qquad\qquad Y_1 \xrightarrow{\ l\ } Y_2$$

is sent to

$$\begin{array}{ll} X_2 \times I \longrightarrow X_1 \times I & I \Rightarrow Y_1 \longrightarrow I \Rightarrow Y_2 \\ (x, i) \longmapsto (k(x), i) & p \longmapsto l \circ p \end{array}$$

respectively.

To show that

$$- \times I \ \dashv\ I \Rightarrow -$$

we at least require an inverse pair of bijections.

$$\begin{array}{c} f \longmapsto \hspace{4cm} f^{\sharp} \\ Set[X, I \Rightarrow Y] \qquad Set[X \times I, Y] \\ g_{\flat} \longleftarrow \hspace{4cm} g \end{array}$$

for each pair X, Y of sets. This is an almost trivial exercise. In many mathematical situations we wouldn't even distinguish between f and $f^{\sharp}$, nor between g and $g_{\flat}$. We should also prove that each of $(\cdot)^{\sharp}$ and $(\cdot)_{\flat}$ is natural. For this example, that is not difficult, but let's leave it until we have more of an understanding of what it entails.

For the record let us state the result we are aiming at.

5.2.1 Theorem *For each set I, we have*

$$- \times I \ \dashv\ I \Rightarrow -$$

an adjunction of endo-functors on Set.

As with any adjunction, this one has a unit and a counit

$$X \xrightarrow{\ \eta_X\ } I \Rightarrow (X \times I) \qquad\qquad (I \Rightarrow Y) \times I \xrightarrow{\ \epsilon_Y\ } Y$$

natural in X and Y, respectively. Here these are more or less obvious.

Exercises

5.2.3 Write down the definitions of $(\cdot)^{\sharp}$ and $(\cdot)_{\flat}$. These two assignments are little more than inserting or omitting brackets.

5.2.4 Write down the unit and counit, and show that each is natural.

5.2.3 A topological example

In this block we re-do the adjunction of Block 5.2.2 with *Set* replaced by *Top*. As we will see, this is not entirely straightforward. We need to impose appropriate conditions on the pivotal object I.

With this kind of result there are two interacting themes, a general and a particular. The general theme is that of the categorical constructions and calculations. Here we find that most of these have been done in Block 5.2.2. That is why we did that simple example before this one. The particular theme is that of handling the topological restrictions. It is useful whenever possible to separate the two themes. Category theory is good at handling the generalities. By separating these from the particularities we see more clearly the specific content of a result.

As in the *Set* example, in this *Top* case there are three players, the central fixed object I and the two varying objects X and Y. Here all three of X, I, Y are topological spaces, and any functor we produce must return topological spaces and continuous maps.

Let

$$(X, \mathcal{O}X) \qquad (I, \mathcal{O}I) \qquad (Y, \mathcal{O}Y)$$

be topological spaces with carried families of open sets, the topologies. We know that the cartesian product $X \times I$ carries the product topology. This is the smallest topology for which the two projections

$$X \times I \longrightarrow X \qquad\qquad X \times I \longrightarrow I$$

are continuous. Thus for each $U \in \mathcal{O}X$ and $W \in \mathcal{O}I$ the product $U \times W$ is open in $X \times I$, and this set of products forms a subbase of the whole topology. This space $X \times I$ with these two continuous projections form a product wedge in *Top*. Thus we have an endo-functor

$$- \times I$$

on *Top*. Our aim is to find a right adjoint to this functor. And we want this to be an enriched hom-functor.

For spaces I and Y let

$$I \Rightarrow Y$$

be the set of all *continuous* maps from I to Y. This is smaller than the set of all functions from I to Y, so perhaps we should use a different notation. But we won't. However, if you do become confused then try a slightly different notation. With I fixed this certainly gives a functor

$$\textbf{\textit{Top}} \xrightarrow{\ I \Rightarrow -\ } \textbf{\textit{Set}}$$

but this isn't good enough. The functor must output to **Top**, not just **Set**. This means we have to find a way of topologizing $(I \Rightarrow Y)$.

Let

$$\mathcal{K}I$$

be the family of compact subsets K of I. Before we continue make sure you know what a compact subset is. Don't go all quasi; it's bad for you!

5.2.2 Definition   For $K \in \mathcal{K}I$ and $V \in \mathcal{O}Y$ let

$$\langle K, V \rangle$$

be the set of continuous maps

$$I \xrightarrow{\ \theta\ } Y \qquad \theta[K] \subseteq V$$

with the indicated property, that is $\theta(i) \in V$ for all $i \in K$. The compact open topology on $(I \Rightarrow Y)$ has the family of all $\langle K, V \rangle$ as a subbase. $\square$

This certainly topologizes $(I \Rightarrow Y)$, but we want $(I \Rightarrow -)$ to be an endofunctor on **Top**, so we also need an action on arrows.

Consider any map ψ between two spaces, as on the left.

$$Y_1 \xrightarrow{\ \psi\ } Y_2 \qquad\qquad (I \Rightarrow Y_1) \xrightarrow{\ \Psi\ } (I \Rightarrow Y_2)$$
$$\theta \longmapsto \psi \circ \theta$$

This induces an assignment Ψ between the two function spaces, as on the right. We check that Ψ is continuous. Consider any $K \in \mathcal{K}I$ and $V \in \mathcal{O}Y_2$. We need

$$\Psi^{\leftarrow}(\langle K, V \rangle)$$

to be open in $I \Rightarrow Y_1$. But a simple calculation gives

$$\Psi^{\leftarrow}(\langle K, V \rangle) = \langle K, \psi^{\leftarrow}(V) \rangle$$

which is open in $(I \Rightarrow Y_1)$ since ψ is continuous. (Do that calculation.)

This gives the arrow assignment for $(I \Rightarrow -)$, and the functorial properties are almost immediate.

We have a pair of endo-functors on *Top*, but these don't necessarily form an adjoint pair. For that we need I to be a particular kind of space.

Read the following definition carefully.

5.2.3 Definition   A topological space I is locally compact if for each situation

$$r \in V \in \mathcal{O}I$$

we have

$$r \in W \subseteq K \subseteq V$$

for some $K \in \mathcal{K}I$ and $W \in \mathcal{O}I$. $\square$

With these preliminaries we have the following. We prove this shortly.

5.2.4 Theorem   *For each locally compact space I, we have*

$$- \times I \dashv I \Rightarrow -$$

an adjunction of endo-functors on **Top**.

Theorem 5.2.4 is important in its own right, but it also has an important refinement. Consider the case where I is the real interval $[0, 1]$, and we modify the two spaces

$$X \times I \qquad\qquad I \Rightarrow Y$$

to produce the

$$\text{suspension space of } X \qquad \text{loop space of } Y$$

respectively. A loop in the space Y is a continuous map

$$\ell : I \longrightarrow Y$$

with $\ell(0) = \ell(1)$. We consider the set of all such loops as a subspace of $I \Rightarrow Y$. We modify the product space $X \times I$ by pinching together all points

$$(x, 0) \quad \text{ for } x \in X$$

and all points

$$(x, 1) \quad \text{ for } x \in X$$

to obtain two pinch points. Technically we take a certain quotient space of $X \times I$. It can be shown that

Suspension $\dashv$ Loop

by adding to the proof of Theorem 5.2.1. (There are some technicalities that I have omitted, but this description is not too far from the truth.)

The proof of Theorem 5.2.4 will take some time. We will do it in little bits as we get to know the general categorical notions. The overall method of proof is the same as for Theorem 5.2.1 with an extra layer of complexity. We must check that various functions are continuous.

Throughout we fix I, an arbitrary locally compact space.

The first thing we must do is set up an inverse pair of assignments

$$
\begin{array}{ccc}
\phi & \longmapsto & \phi^{\sharp} \\
\boldsymbol{Top}[X, I \Rightarrow Y] & & \boldsymbol{Top}[X \times I, Y] \\
\psi_{\flat} & \longleftarrow\!\!\!\dashv & \psi
\end{array}
$$

for arbitrary spaces X and Y. In fact we use the same trick as for the $\boldsymbol{Set}$ case. Thus we set

$$
\psi_{\flat}(x)(i) = \psi(x, i) \qquad \phi^{\sharp}(x, i) = \phi(x)(i)
$$

for each $x \in X$ and $i \in I$. Of course, we must show that $\phi^{\sharp}$ and $\psi_{\flat}$ are continuous, but once we have done that the rest of the proof is trivial (as for the $\boldsymbol{Set}$ case).

We look first at the construction

$$
\psi_{\flat} \longleftarrow\!\!\!\dashv \psi
$$

since this doesn't make use of the local compactness of I. We are given a continuous function

$$
X \times I \xrightarrow{\psi} Y
$$

where $X \times I$ carries the product topology. Since ψ is continuous we see that for each $x \in X$ the function

$$
\psi(x, \cdot) : I \longrightarrow Y
$$

is continuous. Thus we may define a function

$$
\psi_{\flat} : X \longrightarrow (I \Rightarrow Y) \quad \text{by} \quad \psi_{\flat}(x)(i) = \psi(x, i)
$$

for each $x \in X$ and $i \in I$. Our job is to show that $\psi_{\flat}$ is continuous, where $(I \Rightarrow Y)$ carries the compact open topology.

5.2.5 Lemma   *For each pair X and Y of topological spaces, and each continuous map*

$$X \times I \xrightarrow{\;\psi\;} Y$$

the induced function

$$X \xrightarrow{\;\psi_\flat\;} (I \Rightarrow Y)$$

as defined above is continuous.

Proof   Consider any subbasic open set

$$\langle K, V \rangle$$

of $(I \Rightarrow Y)$, where

$$K \in \mathcal{K}I \qquad V \in \mathcal{O}Y$$

are the two components. We require

$$\psi_\flat^\leftarrow (\langle K, V \rangle)$$

to be open in X. Consider any member s of this set. We require

$$s \in U \subseteq \psi_\flat^\leftarrow (\langle K, V \rangle)$$

for some $U \in \mathcal{O}X$. We have

$$\psi_\flat(s) \in \langle K, V \rangle$$

that is

$$\psi(s, i) = \psi_\flat(s)(i) \in V$$

for $i \in K$. We now use the continuity of ψ and the compactness of K.   □

Next we look at the construction

$$\phi \longmapsto \phi^\sharp$$

and this does make use of the local compactness of I.

5.2.6 Lemma   *Let I be a locally compact topological space. For each pair X and Y of topological spaces, and for each continuous map*

$$X \xrightarrow{\;\phi\;} (I \Rightarrow Y)$$

where $(I \Rightarrow Y)$ carries the compact open topology, there is a continuous map

$$X \times I \xrightarrow{\;\phi^\sharp\;} Y$$

given by

$$\phi^\sharp(x, i) = \phi(x)(i)$$

for each $x \in X$ and $i \in I$.

Proof Consider any $V \in \mathcal{O}Y$ and any member

$$(s, r) \in \phi^{\sharp \leftarrow}(V)$$

of the inverse image of V across $\phi^\sharp$. Remembering how $X \times I$ is topologized, it suffices to produce open neighbourhoods

$$s \in U \in \mathcal{O}X \qquad r \in W \in \mathcal{O}I$$

such that

$$U \times W \subseteq \phi^{\sharp \leftarrow}(V)$$

holds. We satisfy the conditions in turn.

We know that

$$\phi(s) : I \longrightarrow Y$$

is continuous, and hence

$$\phi(s)^\leftarrow(V)$$

is open on I. But

$$\phi(s)(r) = \phi^\sharp(s, r) \in V$$

so that

$$r \in \phi(s)^\leftarrow(V)$$

and hence the local compactness of I gives

$$r \in W \subseteq K \subseteq \phi(s)^\leftarrow(V)$$

for some $K \in \mathcal{K}I$ and $W \in \mathcal{O}I$. This W is one of the open sets we need.

The pair K and V give us a subbasic open set

$$\langle K, V \rangle$$

of $I \Rightarrow Y$. Thus, since ϕ is continuous, we see that

$$U = \phi^\leftarrow(\langle K, V \rangle)$$

is open in X. We check that this is the other open set we need. $\qquad \square$

This sets up the inverse pair of bijections for a given pair X, Y. However, we need these bijections to be natural for variation of X and Y. We can postpone a proof of that for a while until we have analysed the general notion in more detail. However, I can tell you that the proof is exactly the same as for the *Set* case. There is no more work to be done.

Exercises

5.2.5 Fill in the details required to show that $(I \Rightarrow -)$ is an endo-functor on *Top*. Many of the calculations are the same as the *Set* case.

5.2.6 Complete the proof of Lemma 5.2.5. As a hint, for a fixed s let i range through K to obtain an open covering of K.

5.2.7 Complete the proof of Lemma 5.2.6. As a hint, observe that for $i \in I$ we have

$$ i \in K \Longrightarrow i \in \phi(s)^{\leftarrow}(V) \Longrightarrow \phi(s)(i) \in V $$

and hence $s \in U$.

5.2.8 For the *Top* adjunction write down the unit and the counit, and show that each is natural.

5.3 Adjunctions uncoupled

In this and the next two sections we look at various aspects of Definition 5.1.1. As we do this you should keep going back to the examples of Section 5.2. This will help you understand the general notions. The *Set* example is always a good place to start, but you should also investigate at least one of the other two examples.

To form an adjunction the data

$$ (F, \ G, \ (\cdot)^{\sharp}, \ (\cdot)_{\flat}) $$

must satisfy two requirements: the bijection requirement and the naturality requirement. The bijection requirement is easy to understand. The naturality requirement needs to be looked at.

The naturality property can be split into several smaller parts. Furthermore, these can be put together in different ways, sometimes in tandom with unit or counit properties, to determine an adjunction. We begin to look at these combinations in this section.

Table 5.1 *The various requirements for an adjunction*

(Bij) For all

Src		*Trg*
A	objects	S
$A \xrightarrow{\ f\ } GS$	arrows	$FA \xrightarrow{\ g\ } S$

from the indicated categories, both

$$(f^\sharp)_\flat = f \qquad (g_\flat)^\sharp = g$$

hold.

(Nat) For all

Src		*Trg*
$B \quad A$	objects	$S \quad T$
$B \xrightarrow{\ k\ } A$		$S \xrightarrow{\ l\ } T$
	arrows	
$A \xrightarrow[\ f\ ]{} GS$		$FA \xrightarrow[\ g\ ]{} S$

from the indicated categories, both

$$(\sharp) \quad (G(l) \circ f \circ k)^\sharp = l \circ f^\sharp \circ F(k) \qquad G(l) \circ g_\flat \circ k = (l \circ g \circ F(k))_\flat \quad (\flat)$$

hold.

The bijection requirement, (Bij), is given in Table 5.1. It merely says that for each pair of objects $A \in$ *Src* and $S \in$ *Trg*, the two assignments

$$
\begin{array}{ccc}
f & \longmapsto & f^\sharp \\
\textit{Src}[A, GS] & & \textit{Trg}[FA, S] \\
g_\flat & \longleftarrow\!\shortmid & g
\end{array}
$$

form an inverse pair of bijections. Each of $(\cdot)^\sharp$ and $(\cdot)_\flat$ determines the other. Thus in any particular example it suffices to mention just one of them, and say that it is a bijection. Its inverse is the other one.

The naturality requirement, (Nat), is more complicated. It says that each of the transposition assignments $(\cdot)^\sharp$ and $(\cdot)_\flat$ is natural. But what does that mean?

Recall that we may form a product category

$$\mathbf{Src} \times \mathbf{Trg}$$

whose objects are pairs

$$(A, S)$$

of objects $A \in \mathbf{Src}$ and $S \in \mathbf{Trg}$. We won't say what the arrows are just yet, and you will see why in a moment. The two functors F and G give functors $\mathfrak{F}$ and $\mathfrak{G}$

$$(A, S) \overset{\mathfrak{F}}{\longmapsto} \mathbf{Trg}[FA, S]$$
$$\mathbf{Src} \times \mathbf{Trg} \longrightarrow \mathbf{Set}$$
$$(A, S) \underset{\mathfrak{G}}{\longmapsto} \mathbf{Src}[A, GS]$$

to the category of sets. Now you can see that we have to be a bit careful. Arrow sets have different variance, namely

$$[\text{Contra} , \text{Co}]$$

in the two positions. Technically we are dealing with a pair of functors

$$\mathbf{Src}^{\text{op}} \times \mathbf{Trg} \overset{\mathfrak{F}}{\underset{\mathfrak{G}}{\rightrightarrows}} \mathbf{Set}$$

where we use the opposite of $\mathbf{Src}$. The naturality requirement (Nat) says that $(\cdot)^{\sharp}$ and $(\cdot)_{\flat}$ provide natural isomorphisms between $\mathfrak{F}$ and $\mathfrak{G}$.

That does describe the requirements, but let's look more closely.

Consider a pair of arrows from the two categories.

$$B \overset{k}{\longrightarrow} A \qquad\qquad S \overset{l}{\longrightarrow} T$$

Notice how we have anticipated the contravariance on the $\mathbf{Src}$ component. The pair (k, l) form an arrow in the product category

$$\mathbf{Src}^{\text{op}} \times \mathbf{Trg}$$

which is the source of both $\mathfrak{F}$ and $\mathfrak{G}$. Using (k, l) we obtain the pair of commuting diagrams in $\mathbf{Set}$ given in Table 5.2. Each arrow

$$A \overset{f}{\longrightarrow} GS \quad \text{of } \mathbf{Src} \qquad\qquad FA \overset{g}{\longrightarrow} S \quad \text{of } \mathbf{Trg}$$

is sent to

$$B \overset{G(l) \circ f \circ k}{\longrightarrow} GT \qquad\qquad FB \overset{l \circ g \circ F(k)}{\longrightarrow} T$$

Table 5.2 Two commuting diagrams in Set

Each pair (k, l) of arrows induces two commuting diagrams

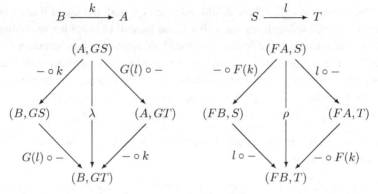

where the components are as follows.

$(A, GS) = \boldsymbol{Src}[A, GS]$	$(FA, S) = \boldsymbol{Trg}[FA, S]$
$(B, GS) = \boldsymbol{Src}[B, GS]$	$(FB, S) = \boldsymbol{Trg}[FB, S]$
$(A, GT) = \boldsymbol{Src}[A, GT]$	$(FA, T) = \boldsymbol{Trg}[FA, T]$
$(B, GT) = \boldsymbol{Src}[B, GT]$	$(FB, T) = \boldsymbol{Trg}[FB, T]$
$\lambda = G(l) \circ - \circ k$	$\rho = l \circ - \circ F(k)$

---

respectively

Now, for given k and l, consider the two paths from $\mathfrak{G}$ to $\mathfrak{F}$.

$$
\begin{array}{ccc}
f & \longmapsto & f\sharp \\
\uparrow & & \uparrow \\
\boldsymbol{Src}[A, GS] \xrightarrow{\;(\cdot)^\sharp\;} \boldsymbol{Trg}[FA, S] & & \\
\downarrow & \downarrow & \\
\boldsymbol{Src}[B, GT] \xrightarrow{\;(\cdot)^\sharp\;} \boldsymbol{Trg}[FB, T] & l \circ f^\sharp \circ F(k) & \\
G(l) \circ f \circ k \longmapsto (G(l) \circ f \circ k)^\sharp & &
\end{array}
$$

The naturality of $(\cdot)^\sharp$ is that these two paths agree for all k, l, f. In the same way, the naturality of $(\cdot)_\flat$ is that the two paths

$$
\begin{array}{ccc}
g_\flat & \longleftarrow & g \\
\uparrow & & \uparrow \\
\boldsymbol{Src}[A, GS] \xleftarrow{\;(\cdot)_\flat\;} \boldsymbol{Trg}[FA, S] & & \\
\downarrow & \downarrow & \\
G(l) \circ g_\flat \circ k \quad \boldsymbol{Src}[B, GT] \xleftarrow{\;(\cdot)_\flat\;} \boldsymbol{Trg}[FB, T] & & l \circ g \circ F(k) \\
(l \circ g \circ F(k))_\flat \longleftarrow & & l \circ g \circ F(k)
\end{array}
$$

agree for all k, l, g.

The naturality requirement, (Nat), is given in Table 5.1. It is stated as two identities

$$(\sharp) \qquad (\flat)$$

in the arrows k, l, f, g. You should note the type and naming of these arrows. We invoke these identities many times, and we can't always use matching letters. And sometimes we need only some of the arrows. In this account we try to stick to the notation used in (Nat). However, sometimes we have to use different letters for the arrows, and sometimes these seem to appear out of position.

By Exercise 3.4.4, if each component of a natural transformation is a bijection, then the inverse is also natural. This simplifies (Nat).

5.3.1 Lemma   *In the presence of* (Bij), *each of the two identities* ($\sharp$), ($\flat$) *of* (Nat) *implies the other.*

Proof   Assuming ($\sharp$) let's check ($\flat$). For this we need arrows k, l, g of the indicated type. Let $f = g_\flat$ to obtain the fourth arrow. From (Bij) we have $g = f^\sharp$. Using ($\sharp$) at the second step we have

$$(G(l) \circ g_\flat \circ k)^\sharp = (G(l) \circ f \circ k)^\sharp = l \circ f^\sharp \circ F(k) = l \circ g \circ F(k)$$

so that

$$G(l) \circ g_\flat \circ k = (l \circ g \circ F(k))_\flat$$

by a second use of (Bij).   □

This shows that in (Nat) only one of the two identities ($\sharp$) and ($\flat$) is needed. So why did we bother to state both? Because some particular examples are best handled using one condition rather than the other. In fact, we can go even further. We can decompose each of ($\sharp$) and ($\flat$) into two bits, and use various combinations of these bits. By taking either k or l to be an identity arrow we obtain four instances of ($\sharp$) and ($\flat$).

$$(\sharp \uparrow) \quad (G(l) \circ f)^\sharp = l \circ f^\sharp \qquad\qquad (\flat \uparrow) \qquad g_\flat \circ k = (g \circ F(k))_\flat$$

$$(\sharp \downarrow) \quad\ \ (f \circ k)^\sharp = f^\sharp \circ F(k) \qquad (\flat \downarrow) \quad G(l) \circ g_\flat = (l \circ g)_\flat$$

In these each occurring arrow has the type given in (Nat). Also, each condition is quantified. For instance, ($\sharp \uparrow$) says

For each pair of arrows

$$A \xrightarrow{\ \ f\ \ } GS \qquad\qquad S \xrightarrow{\ \ l\ \ } T$$

we have ...

and requires only two arrows, as indicated.

The following gives many of the combinations you may meet.

5.3.2 Lemma *Consider a pair of functors*

$$Src \underset{G}{\overset{F}{\rightleftarrows}} Trg$$

and a pair of assignments $(\cdot)^\sharp$ *and* $(\cdot)_\flat$ *satisfying* (Bij). *Then*

$$(\sharp \uparrow) \Longleftrightarrow (\flat \downarrow) \qquad (\sharp \downarrow) \Longleftrightarrow (\flat \uparrow)$$

and the data form an adjunction precisely when one of the four pairs

$$(\sharp \uparrow), (\flat \uparrow)$$
$$(\sharp \uparrow), (\sharp \downarrow) \qquad\qquad (\flat \uparrow), (\flat \downarrow)$$
$$(\sharp \downarrow), (\flat \downarrow)$$

of conditions holds.

Proof The proof of the top two equivalences is more or less the same as that of Lemma 5.3.1. We take

$$k = id_A \qquad l = id_S$$

as appropriate. Similarly, every adjunction satisfies each of the four conditions $(\sharp \uparrow), (\sharp \downarrow), (\flat \uparrow), (\flat \downarrow)$.

It remains to verify that any one of the four pairs ensures that we have an adjunction. Let's look at the pair $(\sharp \uparrow), (\sharp \downarrow)$. It suffices to show

$$(\sharp \uparrow), (\sharp \downarrow) \Longrightarrow (\sharp)$$

and then invoke Lemma 5.3.1. Consider arrows k, l, f as in (Nat). Using $(\sharp \uparrow)$ and then $(\sharp \downarrow)$ we have

$$(G(l) \circ f \circ k)^\sharp = l \circ (f \circ k)^\sharp = l \circ f^\sharp \circ F(k)$$

as required. Notice that $(\sharp \uparrow)$ is applied to the pair

$$B \xrightarrow{f \circ k} GS \qquad\qquad S \xrightarrow{\ l\ } T$$

which is allowed by the quantified condition. $\qquad\qquad\qquad\qquad\qquad$ □

As you can see, there are several combinations of the structured data

$$(F,\ G,\ (\cdot)^\sharp,\ (\cdot)_\flat)$$

which lead to an adjunction. It is best not to try to remember the details, but merely that there are variants.

Exercises

5.3.1   Definition 5.1.1 requires that the bijections $(\cdot)^{\sharp}, (\cdot)_{\flat}$ are natural for variation of both A and S. In Lemma 5.3.1 both A and S vary at the same time. What happens if we fix one of A, S and let the other vary? What conditions arise out of that kind of naturality?

5.3.2   Consider the two algebraic constructions of Block 5.2.1. For both cases show that each of the two assignments $(\cdot)^{\sharp}$ and $(\cdot)_{\flat}$ is natural. In each case draw the square that must commute.

5.3.3   Consider the set-theoretic construction of Block 5.2.2. Show that each of the two assignments $(\cdot)^{\sharp}$ and $(\cdot)_{\flat}$ is natural. In each case draw the square that must commute.

5.3.4   Consider the topological construction of Block 5.2.3. Show that the two assignments $(\cdot)^{\sharp}$ and $(\cdot)_{\flat}$ are natural. In each case draw the square that must commute. Did you notice something about the proof?

5.4 The unit and the counit

In Section 5.1 the unit and the counit of an adjunction made a brief appearance, but were not part of the official gadgetry discussed in Section 5.2. In fact, the unit and counit are important attributes of an adjunction, and can be more important than the transposition assignments. When appropriately restricted they can determine the adjunction.

5.4.1 Definition   Given an adjunction, as in Definition 5.1.1, we set

$$\eta_A = (id_{FA})_{\flat}, \qquad\qquad \epsilon_S = (id_{GS})^{\sharp}$$

for each

$$\textit{Src}\text{-object } A \qquad\qquad \textit{Trg}\text{-object } S$$

to obtain arrows

$$A \xrightarrow{\ \eta_A\ } (G \circ F)A \qquad (F \circ G)S \xrightarrow{\ \epsilon_S\ } S$$

the

$$\text{unit} \qquad\qquad\qquad \text{counit}$$

of the adjunction. (Note that 'counit' is pronounced 'co-unit'.) $\square$

This section is devoted to an analysis of these gadgets. Naturally, we begin with their most important property.

5.4.2 Lemma   *The unit and the counit of an adjunction are natural.*

Proof   We deal with the unit. The counit is dealt with in a similar way.
Consider any arrow

$$B \xrightarrow{\quad k \quad} A$$

of **Src**. We must show that the square

$$
\begin{array}{ccc}
B & \xrightarrow{\eta_B} & (G \circ F)B \\
{\scriptstyle k}\downarrow & & \downarrow {\scriptstyle (G \circ F)(k)} \\
A & \xrightarrow{\eta_A} & (G \circ F)A
\end{array}
\qquad
\begin{array}{l}
\eta_B = (id_{FB})_\flat \\[2em]
\eta_A = (id_{FA})_\flat
\end{array}
$$

commutes, that is we must check that

$$(G \circ F)(k) \circ \eta_B = \eta_A \circ k$$

holds. To do this we use the identity ($\flat$) of (Nat) twice, but in different instantiations.
Thus with

$$B \xrightarrow{\ id_B\ } B \qquad FB \xrightarrow{\ F(k)\ } FA$$

$$FB \xrightarrow[\ id_{FB}\ ]{} FB$$

and then

$$B \xrightarrow{\ k\ } A \qquad FA \xrightarrow{\ id_{FA}\ } FA$$

$$FA \xrightarrow[\ id_{FA}\ ]{} FA$$

we have

$$G(F(k)) \circ (id_{FB})_\flat = (F(k) \circ id_{FB})_\flat = (id_{FA} \circ F(k))_\flat = (id_{FA})_\flat \circ k$$

as required. The central step is a trivial property of identity arrows.   □

Once we know the unit or counit of an adjunction, we can retrieve the transposition assignments.

5.4.3 Lemma   *Given an adjunction, as in Definition 5.1.1, we have*

$$f^{\sharp} = \epsilon_S \circ F(f) \qquad g_{\flat} = G(g) \circ \eta_A$$

for all arrows f and g as in (Nat).

Proof   We verify the left hand equality. Given an arrow

$$A \xrightarrow{\ \ f\ \ } GS$$

we apply ($\sharp \downarrow$) with the following component arrows.

$$(k) \quad A \xrightarrow{\ \ f\ \ } GS \qquad\qquad (f) \quad GS \xrightarrow[id_{GS}]{\ \ } GS$$

Thus

$$f^{\sharp} = (id_{GS} \circ f)^{\sharp} = (id_{GS})^{\sharp} \circ F(f) = \epsilon_S \circ F(f)$$

as required.   □

By remembering Definition 5.4.1 and setting

$$f = \eta_A \qquad g = \epsilon_S$$

we obtain an important particular case of this result.

5.4.4 Corollary   *Given an adjunction, as in Definition 5.1.1, we have*

$$\epsilon_{FA} \circ F(\eta_A) = id_{FA} \qquad G(\epsilon_S) \circ \eta_{GS} = id_{GS}$$

*for each **Src**-object A and **Trg**-object S.*

These are important identities, because in appropriate circumstances they determine the adjunction.

5.4.5 Theorem   *Let*

$$\mathbf{Src} \underset{G}{\overset{F}{\rightleftarrows}} \mathbf{Trg}$$

be a pair of functors, and let

$$Id_{\mathbf{Src}} \xrightarrow{\ \ \eta\ \ } G \circ F \qquad\qquad F \circ G \xrightarrow{\ \ \epsilon\ \ } Id_{\mathbf{Trg}}$$

be natural transformations satisfying the identities of Corollary 5.4.4. Then η and ϵ are the unit and counit of a unique adjunction $F \dashv G$.

Proof By Lemma 5.4.3, if this data does arise from an adjunction then

$$f^\sharp = \epsilon_S \circ F(f) \qquad g_\flat = G(g) \circ \eta_A$$

for all arrows

$$A \xrightarrow{\quad f \quad} GS \qquad\qquad FA \xrightarrow{\quad g \quad} S$$

of *Src* and *Trg*, respectively. Thus it suffices to show that these two assignments $(\cdot)^\sharp, (\cdot)_\flat$ form an inverse pair of bijections which satisfy (Nat).

The given naturality of η ensures that

$$
\begin{array}{ccc}
A & \xrightarrow{\quad \eta_A \quad} & (G \circ F)A \\
{\scriptstyle f}\downarrow & & \downarrow{\scriptstyle G(F(f))} \\
GS & \xrightarrow[\quad \eta_{GS} \quad]{} & (G \circ F \circ G)S
\end{array}
$$

commutes. With this and one of the given conditions we have

$$
\begin{aligned}
(f^\sharp)_\flat &= (\epsilon_S \circ F(f))_\flat & \text{by} && \text{the definition of } (\cdot)^\sharp \\
&= G(\epsilon_S \circ F(f)) \circ \eta_A & \text{by} && \text{the definition of } (\cdot)_\flat \\
&= G(\epsilon_S) \circ G(F(f)) \circ \eta_A & \text{by} && \text{the functorality of } G \\
&= G(\epsilon_S) \circ \eta_{GS} \circ f & \text{by} && \text{the above naturality} \\
&= id_{GS} \circ f & \text{by} && \text{the given right hand identity} \\
&= f & \text{by} && \text{the neutral property of } id_{GS}
\end{aligned}
$$

to give one half of the bijection property. The other half $(g_\flat)^\sharp = g$ is proved in the same way.

The required naturality can be verified in several ways. Let's go for $(\sharp)$. Consider arrows k, l, f as in (Nat). The naturality of ϵ ensures

$$
\begin{array}{ccc}
(F \circ G)S & \xrightarrow{\quad \epsilon_S \quad} & S \\
{\scriptstyle F(G(l))}\downarrow & & \downarrow{\scriptstyle l} \\
(F \circ G)T & \xrightarrow[\quad \epsilon_T \quad]{} & T
\end{array}
$$

commutes. Using this we have

$$
\begin{aligned}
l \circ f^\sharp \circ F(k) &= l \circ \epsilon_S \circ F(f) \circ F(k) & \text{by} && \text{the definition of } (\cdot)^\sharp \\
&= l \circ \epsilon_S \circ F(f \circ k) & \text{by} && \text{the functorality of } F \\
&= \epsilon_T \circ F(G(l)) \circ F(f \circ k) & \text{by} && \text{the naturality from above} \\
&= \epsilon_T \circ F(G(l) \circ f \circ k) & \text{by} && \text{the functorality of } F \\
&= (G(l) \circ f \circ k)^\sharp & \text{by} && \text{the definition of } (\cdot)^\sharp
\end{aligned}
$$

to give $(\sharp)$. The identity $(\flat)$ can be verified in the same way. $\qquad\square$

For any particular adjunction only some of the data

$$(F, \ G, \ (\cdot)^{\sharp}, \ (\cdot)_{\flat}, \ \eta, \ \epsilon)$$

is needed. However, given an adjunction you should get into the habit of working out what each component is.

Exercises

5.4.1 Show that the counit of an adjunction is natural.

5.4.2 Show how to retrieve the transposition $(\cdot)_{\flat}$ from the unit η.

5.4.3 Do the other half of the proof of Theorem 5.4.5.

5.4.4 Consider the two algebraic constructions of Block 5.2.1.
  Show that for each set X and algebra A there are assignments

$$X \xrightarrow{\ \eta_X\ } (U \circ \Sigma)X \qquad (\Sigma \circ U)A \xrightarrow{\ \delta_A\ } A$$

$$(U \circ \Pi)X \xrightarrow{\ \epsilon_X\ } X \qquad A \xrightarrow{\ \zeta_A\ } (\Pi \circ U)A$$

where δ_A and ζ_A are morphisms.
  Show that each of the families $\eta, \epsilon, \delta, \zeta$ is a natural transformation.

5.4.5 For the two algebraic constructions of Block 5.2.1, verify directly the identities of Lemma 5.4.3 and Corollary 5.4.4. In each case draw the composite arrow that is dealt with.

5.4.6 For the set-theoretic and topological constructions of Blocks 5.2.2 and 5.2.3, verify directly the identities of Lemma 5.4.3 and Corollary 5.4.4. In each case draw the composite arrow that is dealt with.

5.5 Free and cofree constructions

Adjunctions can arise in several different guises and even disguises, and some of these don't look at all like the official notion. These differences are probably the reason why the general notion wasn't recognized earlier. In this section we look at two such examples. You will recognize the notions from earlier. They are the idea of the universal

free   cofree

solution across a functor which now need not be forgetful. (Note that 'cofree' is pronounced 'co-free'.)

I am going to describe the two notions in parallel. I suggest that you read just one side first, and perhaps the free side is easier. Once you are almost happy with that, go through the other side. You should note the symmetry between the two sides. At the first reading do not try to connect the ideas with that of an adjunction, even though some of the notation is the same. We will sort that out later.

For both notions we again have a pair

$$Src \qquad Trg$$

of categories. But now we have just one functor

free   cofree

$$Src \xleftarrow{\quad G \quad} Trg \qquad\qquad Src \xrightarrow{\quad F \quad} Trg$$

depending on which side we are considering, as indicated. Often in particular examples this is a forgetful functor, but it need not be.

The idea is that we want to convert each

$$Src\text{-object } A \qquad\qquad Trg\text{-object } S$$

into an object of the other category. Furthermore, we want to do this in the most economical fashion. Thus we pose two problems, the

free problem   cofree problem

respectively. (In fact, we don't say what the problem is, we merely say what a solution is.)

For the problem a solution is an arrow

$$A \xrightarrow{\quad f \quad} GS \qquad\qquad FA \xrightarrow{\quad g \quad} S$$

comparing a

$$Src\text{-object } A \qquad\qquad Trg\text{-object } S$$

with a transposed

$$Trg\text{-object } S \qquad\qquad Src\text{-object } A$$

respectively. Note the direction of the comparison. It is from *Src* to *Trg* in both problems. The difference between the two problems is the object that is transported to the other side.

We look for a universal solution to the problem which applies to each

$$Src\text{-object } A \qquad\qquad Trg\text{-object } S$$

respectively. We might think of this as a uniform universal solution. Thus we look for an object assignment

Src	Trg	Src	Trg
$A \longmapsto FA$		$GS \longmapsfrom S$	

together with a selected arrow

$$A \xrightarrow{\eta_A} (G \circ F)A \qquad (F \circ G)S \xrightarrow{\epsilon_S} S$$

for each object. Observe that, as yet, part of the solution is just an object assignment, not a functor. (Eventually we show that it is a functor, but that is not part of the required solution.)

So far these gadgets merely select a solution to the problem. We want a *universal* solution, a solution through which each other solution must factor uniquely.

5.5.1 Definition   The notions of a

$$\text{free} \qquad\qquad \text{cofree}$$

solution are defined in unison as in Table 5.3. You should read each column separately, perhaps starting with the free (left hand) column.   □

This definition looks quite complicated. The way to remember it is

For each f there is an $f^\sharp$ such that the triangle commutes.	For each g there is a $g_\flat$ such that the triangle commutes.

and then let the parent object A or S vary.

In due course we will see that having a free or cofree solution is equivalent to the given functor having an adjoint on the appropriate side. In fact, sometimes Definition 5.5.1 is taken as the official definition of an adjunction, and that is certainly a useful way of remembering some of the facets of adjunctions. However, there are certain generalizations of the notion of an adjunction (where the *sets* of arrows are given some other structure). In those circumstances the free/cofree version doesn't work so well.

In this section we do two things. We show first how an adjunction gives a free and a cofree solution. Then we show how every free and cofree solution arises from an adjunction.

Table 5.3 *Free and Cofree solutions*

Free	Cofree
Let	Let

Free

Let

$$Src \xleftarrow{\quad G \quad} Trg$$

be a functor and consider

$$A \longmapsto FA$$

an object assignment in the other direction.

A *Src*-indexed family

$$A \xrightarrow{\eta_A} (G \circ F)A$$

of *Src*-arrows forms a

$$G\text{-free}$$

solution if for each *Src*-arrow

$$A \xrightarrow{f} GS$$

with $S \in Trg$, there is a *unique* *Trg*-arrow

$$FA \xrightarrow{f^\sharp} S$$

such that

$$A \xrightarrow{f} GS$$
$$\eta_A \searrow \quad \nearrow G(f^\sharp)$$
$$(G \circ F)A$$

commutes.

Cofree

Let

$$Src \xrightarrow{\quad F \quad} Trg$$

be a functor and consider

$$GS \longleftarrow\!\mid S$$

an object assignment in the other direction.

A *Trg*-indexed family

$$(F \circ G)S \xrightarrow{\epsilon_S} S$$

of *Trg*-arrows form a

$$F\text{-cofree}$$

solution if for each *Trg*-arrow

$$FA \xrightarrow{g} S$$

with $A \in Src$, there is a *unique* *Src*-arrow

$$A \xrightarrow{g_\flat} GS$$

such that

$$FA \xrightarrow{g} S$$
$$F(g_\flat) \searrow \quad \nearrow \epsilon_S$$
$$(F \circ G)S$$

commutes.

5.5.2 Theorem   *Let*

$$Src \quad \underset{G}{\overset{F}{\rightleftarrows}} \dashv \quad Trg$$

be an adjunction with associated gadgets in standard notation.

The object assignment F, the unit η, and the transposition $(\cdot)^\sharp$ provide the data for a G-free solution.

The object assignment G, the counit ϵ, and the transposition $(\cdot)_\flat$ provide the data for an F-cofree solution.

Proof   We look at the free result.

Consider any arrow

$$A \xrightarrow{\ f\ } GS$$

of *Src*. We first check that

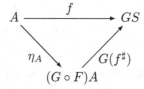

does commute (and then consider the required uniqueness). We use the selection of arrows

$$A \xrightarrow{\ id_A\ } A \qquad\qquad FA \xrightarrow{\ f^\sharp\ } S$$

$$FA \xrightarrow[id_{FA}]{} FA$$

and then apply ($\flat$) of (Nat). Thus

$$G(f^\sharp) \circ \eta_A = G(f^\sharp) \circ (id_{FA})_\flat \circ id_A = (f^\sharp \circ id_{FA} \circ F(id_A))_\flat = (f^\sharp)_\flat = f$$

as required.

For the uniqueness we consider any arrow

$$FA \xrightarrow{\ g\ } S \quad \text{for which} \quad f = G(g) \circ \eta_A$$

and show that, in fact, $g = f^\sharp$. We use the selection of arrows

$$A \xrightarrow{\ id_A\ } A \qquad\qquad FA \xrightarrow{\ g\ } S$$

$$FA \xrightarrow[id_{FA}]{} FA$$

and then apply ($\flat$) of (Nat). Thus

$$f = G(g) \circ \eta_A$$
$$= G(g) \circ (id_{FA})_\flat$$
$$= G(g) \circ (id_{FA})_\flat \circ id_A$$
$$= (g \circ id_{FA} \circ F(id_A))_\flat = g_\flat$$

and hence

$$g = (g_\flat)^\sharp = f^\sharp$$

by a use of (Bij). □

This shows that each adjunction gives universal solutions of both parities. The more important result is that every universal solution must arise from an adjunction.

5.5.3 Theorem   *Let*

$$Src \xleftarrow{\quad G \quad} Trg$$

be a functor, and suppose

$$F \quad \eta \quad (\cdot)^\sharp$$

is the data that provides a G-free solution. Then the object assignment F fills out to a functor for which

$$F \dashv G$$

with $(\cdot)^\sharp$ as the transposition assignment and η as the unit.

Proof   The proof is quite long, but not very deep. There are many small parts each of which is straightforward. The G-free solution says

> For each arrow f (of a certain kind), there is a
> *unique* arrow $f^\sharp$ (to do a certain job).

and it is this uniqueness that is important. We use it some 8 or 9 times.

We must first produce an arrow assignment to create the functor F. Consider any arrow

$$B \xrightarrow{\quad k \quad} A$$

of *Src*, so that an arrow

$$FB \xrightarrow{\quad F(k) \quad} FA$$

is required together with some appropriate properties.

Let f be the composite

$$B \xrightarrow{\;k\;} A \xrightarrow{\;\eta_A\;} (G \circ F)A$$

and consider the commuting square

$$
\begin{array}{ccc}
B & \xrightarrow{\;k\;} & A \\
\eta_B \downarrow & \quad (\square) & \downarrow \eta_A \\
(G \circ F)B & \xrightarrow[G(f^\sharp)]{} & (G \circ F)A
\end{array}
$$

provided by the G-free solution. Here we have

$$FB \xrightarrow{\;f^\sharp\;} FA$$

and we take this to be $F(k)$. Thus, we set

$$F(k) = (\eta \circ k)^\sharp$$

for each **Src**-arrow k, as above.

This gives an arrow assignment. We show that F is a functor. We check that F passes across composition, and preserves identity arrows.

Consider a composite

$$C \xrightarrow{\;l\;} B \xrightarrow{\;k\;} A \quad \text{with} \quad m = k \circ l$$

in **Src**. We require

$$F(m) = F(k) \circ F(l)$$

in **Trg**. Remember what $F(m)$ is. It is the *unique* arrow

$$FC \xrightarrow{\qquad} FA$$

such that

$$
\begin{array}{ccc}
C & \xrightarrow{\;m\;} & A \\
\eta_C \downarrow & & \downarrow \eta_A \\
(G \circ F)C & \xrightarrow[G(F(m))]{} & G(\circ F)A
\end{array}
$$

commutes. But, by construction of $F(k)$ and $F(l)$, both of the squares

$$
\begin{array}{ccccc}
C & \xrightarrow{\ \ l\ \ } & B & \xrightarrow{\ \ k\ \ } & A \\
\eta_C \downarrow & & \eta_B \downarrow & & \downarrow \eta_A \\
(G \circ F)C & \xrightarrow[G(F(l))]{} & (G \circ F)B & \xrightarrow[G(F(k))]{} & (G \circ F)A
\end{array}
$$

commute, and hence

$$
\begin{array}{ccc}
C & \xrightarrow{\hspace{4cm} m \hspace{4cm}} & A \\
\eta_C \downarrow & & \downarrow \eta_A \\
(G \circ F)C & \xrightarrow[G(F(k) \circ F(l)) = G(F(k)) \circ G(F(l))]{} & G(\circ F)A
\end{array}
$$

commutes (for we know that G is a functor). The *uniqueness* now gives

$$ F(k \circ l) = F(m) = F(k) \circ F(l) $$

which is what we want.

We must check that F preserves identity arrows, that is

$$ F(id_A) = id_{FA} $$

for each object A of *Src*. But $F(id_A)$ is the *unique* arrow

$$ FA \xrightarrow{\ \ g\ \ } FA $$

such that

$$
\begin{array}{ccc}
A & \xrightarrow{\ \ id_A\ \ } & A \\
\eta_A \downarrow & & \downarrow \eta_A \\
(G \circ F)A & \xrightarrow[G(g)]{} & (G \circ F)A
\end{array}
$$

commutes. Since

$$ G(id_{FA}) = id_{(G \circ F)A} $$

we see that

$$ g = id_{FA} $$

does make this latest square commute, and hence

$$ F(id_A) = g = id_{FA} $$

by yet another appeal to the *uniqueness*.

This gives us a functor F, and the commuting square ($\square$) shows that η is natural. We now check that

$$F \dashv G$$

using the assignment $(\cdot)^\sharp$.

We look first at (Bij). Fix $A \in \textbf{\textit{Src}}$ and $S \in \textbf{\textit{Trg}}$. We certainly have an assignment

$$\textbf{\textit{Src}}[A, GS] \longrightarrow \textbf{\textit{Trg}}[FA, S]$$
$$f \longmapsto f^\sharp$$

between the indicated arrow sets. We show that this is a bijection (and then $(\cdot)_\flat$ is its inverse). By definition of G-free, for each arrow

$$A \xrightarrow{\ f\ } GS$$

the arrow

$$FA \xrightarrow{\ g = f^\sharp\ } S$$

must ensure that

commutes, and is the *unique* arrow to do this.

To show that $(\cdot)^\sharp$ is injective, suppose

$$f_1^\sharp = f_2^\sharp$$

for two arrows taken from $\textbf{\textit{Src}}[A, GS]$. Then

$$f_1 = G(f_1^\sharp) \circ \eta_A = G(f_1^\sharp) \circ \eta_A = f_2$$

as required. To show that $(\cdot)^\sharp$ is surjective, consider any arrow g taken from $\textbf{\textit{Trg}}[FA, S]$. Let

$$f = G(g) \circ \eta_A$$

so that ($\triangledown$) commutes. But now, by the *uniqueness* we have

$$f^\sharp = g$$

to give the required result.

Next we verify (Nat). We already have (Bij) so it suffices to check ($\sharp$). To this end consider arrows

$$B \xrightarrow{\ k\ } A \qquad\qquad S \xrightarrow{\ l\ } T$$

$$A \xrightarrow[\ f\ ]{} GS$$

and let

$$h = (G(l) \circ f \circ k)^{\sharp}$$

so that

$$h = l \circ f^{\sharp} \circ F(k)$$

is our problem. By construction, h is the *unique* arrow

$$FB \xrightarrow{\ h\ } T$$

such that

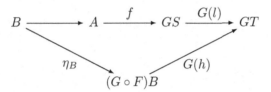

commutes. Now consider the diagram

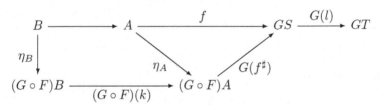

and observe that the left hand square and central triangle do commute. Since G is a functor, this gives a commuting triangle

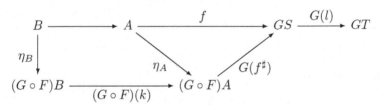

and hence once again the *uniqueness* is the answer to our problem.

Don't go away just yet, for we haven't quite finished. Read the statement of the result again. This says that under the given circumstances we have

$$F \dashv G$$

with $(\cdot)^\sharp$ as the transposition assignment *and* with η as the unit. We still have this last clause to verify.

We require

$$\eta_A = (id_{FA})_\flat$$

for each object A of **Src**. With $S = FA$, consider the arrow

$$A \xrightarrow{\ \eta_A\ } GS$$

of **Src**. What is the job done by $\eta^\sharp$. It is the *unique* arrow

$$FA \xrightarrow{\ g\ } S$$

such that

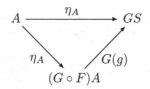

commutes. But clearly, since

$$G(id_{FA}) = id_{GS}$$

we see that

$$g = id_{FA}$$

does this job, and hence

$$\eta_A^\sharp = id_{FA}$$

by a final appeal to *uniqueness*. Since $(\cdot)_\flat$ is the inverse of $(\cdot)^\sharp$, we have

$$\eta_A = (\eta_A^\sharp)_\flat = (id_{FA})_\flat$$

as required. $\square$

Now you can relax, but not too much. You have to do the cofree proof.

Exercises

5.5.1 Finish the proof of Theorem 5.5.2. That is, do the cofree part.

5.5.2 Do the cofree analogue of Theorem 5.5.3.

5.5.3 In this exercise the forgetful functor has been omitted. You should insert it where necessary.

Consider the algebraic construction of Block 5.2.1. Using the object assignment $X \longmapsto \Sigma X$ given there, and the function

$$X \xrightarrow{\quad \eta_X \quad} \Sigma X$$

given by Exercise 5.4.4, show that for each function

$$X \xrightarrow{\quad f \quad} A$$

from a set to an algebra, there is a unique morphism

$$\Sigma X \xrightarrow{\quad f^\sharp \quad} A$$

such that the following commutes in *Set*.

$$X \xrightarrow{\quad f \quad} A$$
$$\eta_X \searrow \quad \nearrow f^\sharp$$
$$\Sigma X$$

This shows that ΣX is the free algebra generated by X, via η_X.

5.5.4 In this exercise the forgetful functor has been omitted. You should insert it where necessary.

Consider the algebraic construction of Block 5.2.1. Using the object assignment $X \longmapsto \Pi X$ given there, and the function

$$\Pi X \xrightarrow{\quad \epsilon_X \quad} X$$

given by Exercise 5.4.4, show that for each function

$$A \xrightarrow{\quad g \quad} X$$

from an algebra to a set, there is a unique morphism

$$A \xrightarrow{\quad g_\flat \quad} \Pi X$$

such that the following commutes in *Set*.

$$
\begin{array}{ccc}
A & \xrightarrow{\quad f \quad} & X \\
 & {}_{g_b}\searrow \quad \nearrow_{\epsilon_X} & \\
 & \Pi X &
\end{array}
$$

This shows that ΠX is the cofree algebra co-generated by X, via ϵ_X.

5.5.5 Verify the free and cofree properties for the constructions of Blocks 5.2.2 and 5.2.3.

5.6 Contravariant adjunctions

We have looked at adjoint pairs of covariant functors. There is a similar notion for contravariant functors. In some ways this is easier because the notions are completely symmetric between the two component functors.

5.6.1 Definition   Let

$$
Alg \xrightarrow{\ \mathfrak{S}\ } Spc \qquad\qquad Alg \xleftarrow{\ \mathfrak{A}\ } Spc
$$

be a pair of contravariant functors between a pair of categories These form a contravariant adjunction if for each

$$
\textit{Alg}\text{-object } A \qquad \textit{Spc}\text{-object } S
$$

there is a bijective correspondence

$$
\begin{array}{cc}
Alg[A, \mathfrak{A}S] & Spc[S, \mathfrak{S}A] \\
f \longmapsto & f^{\sigma} \\
\phi^{\alpha} \longleftarrow\!\!\!\!\!-\!\!\!\!- & \phi
\end{array}
$$

between the two arrow sets. Furthermore, this correspondence must be natural for variation of A and S.   □

The notation here is chosen to be suggestive. The two categories

$$
\textit{Alg} \qquad\qquad \textit{Spc}
$$

are often of an

$$
\text{algebraic} \qquad\qquad \text{spatial}
$$

nature. Each of the two functors

$$
\mathfrak{A} \qquad\qquad \mathfrak{S}
$$

is named after its target.

As with a covariant adjunction, each identity arrow

$$\mathfrak{A}S \xrightarrow{\ id_{\mathfrak{A}S}\ } \mathfrak{A}S \qquad\qquad \mathfrak{S}A \xrightarrow{\ id_{\mathfrak{S}A}\ } \mathfrak{S}A$$

can be transferred to the other side to produce arrows

$$A \xrightarrow{\ h\ } (\mathfrak{A}\circ\mathfrak{S})A \qquad\qquad S \xrightarrow{\ \eta\ } (\mathfrak{S}\circ\mathfrak{A})S$$

the analogues of the unit. Often one or both of these form a representation of the parent object in terms of a gadget of the other kind. At the heart of many representation results there is a contravariant adjunction. Exercise 5.6.2 looks at some of the details of this.

The required naturality is worth looking at. Consider a pair of arrows

$$B \xrightarrow{\ l\ } A \qquad\qquad T \xrightarrow{\ \lambda\ } S$$

from the two categories. These induce a square

between the arrow sets, which must commute. Here, for arrows f and ϕ as indicated, we have

$$f^{\bullet} = \mathfrak{A}(\lambda) \circ f \circ l \qquad \phi^{\bullet} = \mathfrak{S}(l) \circ \phi \circ \lambda$$

and we require

$$f^{\bullet\alpha} = f^{\sigma\bullet} \qquad \phi^{\alpha\bullet} = \phi^{\bullet\sigma}$$

to hold.

Exercises

5.6.1 In this exercise we first set up two contravariant functors

$$Pos \xrightarrow{\ \Upsilon\ } Top \qquad\qquad Pos \xleftarrow{\ \mathcal{O}\ } Top$$

between the category of posets and monotone functions and the category of topological spaces and continuous maps. We then show that these form a contravariant adjunction.

(a) The functor $\mathcal{O}$ views the topology $\mathcal{O}S$ of a space S as a poset. Where have you seen this functor before?

(b) For each poset A let ΥA be the family of upper sections of A. For each finite subset a of A let $\langle a \rangle$ be the subset of ΥA given by

$$p \in \langle a \rangle \iff a \subseteq p$$

for $p \in \Upsilon A$. Show that these subsets form a base for a topology on ΥA.

(c) Show that for each monotone function

$$A \xrightarrow{\quad f \quad} B$$

between posets, the inverse image map

$$\Upsilon B \xrightarrow{\ \phi = f^{\leftarrow}\ } \Upsilon A$$

is continuous (relative to the carried topologies), and hence Υ is a contravariant functor.

(d) Show that the two functors form a contravariant adjunction.

(e) Describe the two units and show that each is an arrow of the appropriate category.

(f) Can you see a neater way of setting up this adjunction?

5.6.2   In this exercise we look at the idea of a schizophrenically induced contravariant adjunction. Not all the details are dealt with, so we take some on trust.

Suppose we have two *Set*-based categories.

$$Alg \qquad Spc$$

Thus each object is a furnished set and each arrow is a function of a certain kind. In many examples one has an algebraic nature and the other has a spatial nature, hence the notation used here. Suppose we have a gadget ★ which lives in both categories. That is, we have a set that can be furnished in two ways to produce an *Alg*-object or a *Spc*-object. This is the schizophrenic object.

For each

$$A \in Alg \qquad S \in Spc$$

we have arrow sets

$$Alg[A, \bigstar] \qquad Spc[S, \bigstar]$$

in other words we have contravariant hom-functors

$$Alg \xrightarrow{\qquad} Set \qquad\qquad Spc \xrightarrow{\qquad} Set$$

where the behaviour on arrows is via composition. The nature of ★ enables us to enrich

$$\mathfrak{G}A = \textit{Alg}[A, \bigstar] \qquad \mathfrak{A}S = \textit{Spc}[S, \bigstar]$$

so that they are objects of

$$\textit{Spc} \qquad\qquad \textit{Alg}$$

respectively. This step is not routine; it is concerned with the non-categorical aspects of the situation under analysis. We don't worry about the details here. When the construction works these enrichments are compatible with composition, to give a pair of contravariant functors

$$\textit{Alg} \xrightarrow{\;\mathfrak{G}\;} \textit{Spc} \qquad\qquad \textit{Spc} \xrightarrow{\;\mathfrak{A}\;} \textit{Alg}$$

between the categories. The functorial behaviour is simply that of a hom-functor. The enrichments are carried along almost without effort.

This exercise shows these functors form a contravariant adjunction.

(a) Show that for each pair of objects $A \in \textit{Alg}$ and $S \in \textit{Spc}$ there is a bijective correspondence

$$\textit{Alg}[A, \mathfrak{A}S] \qquad \textit{Spc}[S, \mathfrak{G}A]$$
$$f \longleftrightarrow \phi$$

between the two arrow sets. (To do this first curry the function types and then chip the inputs, but don't make a meal of it.)

(b) Show that this correspondence is natural for variation of A and S. You need only worry about the *Set* aspects.

(c) Write down the two units

$$A \xrightarrow{\;h_A\;} (\mathfrak{A} \circ \mathfrak{G})A \qquad\qquad S \xrightarrow{\;\eta_S\;} (\mathfrak{G} \circ \mathfrak{A})S$$

for arbitrary $A \in \textit{Alg}$ and $S \in \textit{Spc}$.

6

Posets and monoid sets

In this last chapter we look at some examples of adjunctions. These are divided into two groups. In Sections 6.1 to 6.6 we look at two ways that a poset can be completed. In Sections 6.7 to 6.11 we look at two ways that a set can be converted into an R-set for a given monoid R. For both groups most of the details are left as exercises.

Finally, in Section 6.12 several small projects are suggested. Each is to modify one of the various adjunctions discussed earlier. All the answers are known, but some of them are not as well-known as they should be.

6.1 Posets and complete posets

Let Pos be the category of posets and monotone maps. We have used this on several occasions to illustrate various notions. We are concerned here also with the class of complete posets, those which have all suprema and all infima. At first sight these appear to form a subcategory of Pos, but a closer look shows they are the objects of *two* categories. We need to take some care with the comparison arrows between complete posets. Each of these two subcategories is reflective in Pos. We look at the details of these two reflections.

Exercises

6.1.1 Let

$$S \xrightarrow{\quad f \quad} T$$

be a monotone map between two complete posets. Find a necessary and sufficient condition that f has a right adjoint. What about the existence of a left adjoint?

6.2 Two categories of complete posets

Let's collect together the various standard notions that we need.

6.2.1 Definition   Le S be an arbitrary poset, and let X be an arbitrary subset of S. A

upper   lower

bound for X in S is an element $a \in S$ such that

$$x \le a \qquad a \le x$$

for each $x \in X$.

The

supremum $\bigvee X$ infimum $\bigwedge X$

of X in S is an upper bound a such that

$$a \le b \qquad b \le a$$

for each other

upper   lower

bound of X. In other words, it is the

least upper   greatest lower

bound of X in S. $\qquad\qquad\qquad\qquad\qquad\qquad\qquad\qquad\qquad\qquad$ □

Of course, any given $X \subseteq S$ need not have an upper bound in S. Even if it does, it need not have a supremum in S. Similarly, it need not have a lower bound, and even if it does, it need not have an infimum. Thus the existence of such elements imposes a restriction on the poset S.

6.2.2 Definition   A poset S is

$\bigvee$-complete $\bigwedge$-complete

if

$$\bigvee X \qquad\qquad \bigwedge X$$

exists for each subset X of S. $\qquad\qquad\qquad\qquad\qquad\qquad\qquad\qquad$ □

The following simple observation is sometimes a surprise.

6.2.3 Lemma   *A poset S is $\bigvee$-complete if and only if it is $\bigwedge$-complete.*

The two properties of Definition 6.2.2 are equivalent. So why don't we simply refer to a **complete poset** rather than try to distinguish between two kinds? That is the common practice, and we will follow that here. However, when we try to make a category out of complete posets we find there are at least two possibilities. For those categories we need the terminology of Definition 6.2.2.

6.2.4 Definition   Let S, T be a pair of complete posets. A

$$\bigvee\text{-morphism} \qquad \bigwedge\text{-morphism}$$

between the two is a monotone map

$$S \xrightarrow{\ f\ } T$$

such that

$$f(\bigvee X) = \bigvee f[X] \qquad f(\bigwedge X) = \bigwedge f[X]$$

for each $X \subseteq S$.   □

Here $f[\cdot]$ indicates the direct image function across f. Thus

$$f(\bigvee X) = \bigvee \{f(x) \mid x \in X\} \qquad f(\bigwedge X) = \bigwedge \{f(x) \mid x \in X\}$$

are the two extra restrictions on the monotone function f. However, you should note that a $\bigvee$-morphism need not be a $\bigwedge$-morphism, and conversely.

Almost trivially, these respective morphisms are closed under composition, so the following makes sense.

6.2.5 Definition   Complete posets are the objects of the two categories

$$\textit{Sup} \qquad \textit{Inf}$$

and the arrows are the

$$\bigvee\text{-morphisms} \qquad \bigwedge\text{-morphisms}$$

respectively.   □

Trivially, we have a pair of forgetful functors

$$\textit{Pos} \longleftarrow \textit{Sup} \qquad \textit{Pos} \longleftarrow \textit{Inf}$$

and we aim to produce a left adjoint for each of these. Thus we require two kinds of completion processes for posets.

Exercises

6.2.1   When does

$$\bigvee \emptyset \quad \bigvee S \qquad \bigwedge \emptyset \quad \bigwedge S$$

exist in a poset S, and in each case what is the element?

6.2.2   Find examples of a poset S and a subset X where X has

an upper   a lower

bound but

$$\bigvee X \qquad \bigwedge X$$

does not exist. Can you find such examples in a finite poset?

6.2.3   Prove Lemma 6.2.3.

6.2.4   Find an example of a $\bigvee$-morphism that is not a $\bigwedge$-morphism.
  Find an example of a monotone map between complete posets that is not a $\bigvee$-morphism.

6.3 Sections of a poset

To complete a poset we use certain of its subsets.

6.3.1 Definition   Let S be an arbitrary poset. A subset $X \subseteq S$ is

a lower section   an upper section

of S if,

$$\left.\begin{array}{c} x \in X \\ y \le x \end{array}\right\} \Longrightarrow y \in X \qquad \left.\begin{array}{c} x \in X \\ x \le y \end{array}\right\} \Longrightarrow y \in X$$

for all $x, y \in S$.   □

  Observe that

$$X \text{ lower} \Longrightarrow X' \text{ upper} \qquad X \text{ upper} \Longrightarrow X' \text{ lower}$$

where $(\,\cdot\,)'$ produces the complement in S. Thus there is a bijective correspondence between the lower sections and the upper sections of S.

6.3.2 Definition Let S be an arbitrary poset. For each $X \subseteq S$ we let

$$\downarrow X \qquad \uparrow X$$

be the

$$\text{lower section} \qquad \text{upper section}$$

generated by X. Thus

$$y \in \downarrow X \Longleftrightarrow (\exists x \in X)[y \le x] \qquad y \in \uparrow X \Longleftrightarrow (\exists x \in X)[x \le y]$$

for each $y \in S$.

We also let

$$\downarrow a = \downarrow\{a\} \qquad \uparrow a = \uparrow\{a\}$$

be the principal

$$\text{lower} \qquad \text{upper}$$

section generated by the element $a \in S$. Thus

$$y \in \downarrow a \Longleftrightarrow y \le a \qquad y \in \uparrow a \Longleftrightarrow a \le y$$

for each $y \in S$. $\qquad\qquad\square$

The family of all lower sections of S is a poset under inclusion. We use this poset as the object assignment of at least two functors.

6.3.3 Definition For each poset S we let $\mathcal{L}S$ be the poset of all lower sections of S under inclusion. $\qquad\qquad\square$

It is easy to show that $\mathcal{L}S$ is closed under arbitrary unions and intersection, and so $\mathcal{L}S$ is a complete poset. We use this to form two completions of S.

6.3.4 Definition For each poset S and element $a \in S$ we set

$$\eta_S^{\exists}(a) = \downarrow a \qquad \eta_S^{\forall}(a) = (\uparrow a)'$$

to obtain the assignments

$$S \xrightarrow{\ \eta_S^{\exists}\ } \mathcal{L}S \qquad\qquad S \xrightarrow{\ \eta_S^{\forall}\ } \mathcal{L}S$$

respectively. $\qquad\qquad\square$

It is easy to show that each of $\eta_S^{\exists}$ and $\eta_S^{\forall}$ is monotone, although the second one does require just a little bit more thought.

Exercises

6.3.1 Show the following for an arbitrary poset.
The union of a family of lower sections is a lower section.
Each lower section is the union of a family of principal lower sections.
The intersection of a family of lower sections is a lower section.

6.3.2 Show that each of $\eta_S^{\exists}$ and $\eta_S^{\vee}$ is monotone.
What about the assignment $a \longmapsto \uparrow a$?

6.4 The two completions

For each poset S the associated poset $\mathcal{L}S$ of lower sections is closed under
arbitrary unions and intersection, and so is a complete poset. Also S is con-
nected to $\mathcal{L}S$ by two different monotone maps $\eta^{\exists}$ and $\eta^{\vee}$. We investigate the
properties of these maps, and in due course we show they are reflecting maps.

6.4.1 Lemma *Let S be an arbitrary poset. We have*

$$\downarrow X = \bigcup \eta^{\exists}[X] \qquad (\uparrow X)' = \bigcap \eta^{\vee}[X]$$

for each subset $X \subseteq S$.

This ensures that $\eta^{\exists}, \eta^{\vee}$ are 'sufficiently epic' in the following sense.

6.4.2 Lemma *For each poset S and each parallel pair*

$$\mathcal{L}S \underset{h}{\overset{g}{\rightrightarrows}} T$$

of

$$\bigvee\text{-morphisms} \qquad \bigwedge\text{-morphisms}$$

to a complete poset T, we have

$$g \circ \eta^{\exists} = h \circ \eta^{\exists} \implies g = h \qquad g \circ \eta^{\vee} = h \circ \eta^{\vee} \implies g = h$$

respectively.

It is now easy to obtain the two completions. Notice how the following gives
two versions of 'freely generated by' a poset.

6.4.3 Theorem   *For each poset S and monotone map*

$$S \xrightarrow{\quad f \quad} T$$

to a complete poset T, there is a unique

$$\bigvee\text{-morphism} \qquad \bigwedge\text{-morphism}$$

$$\mathcal{L}S \xrightarrow{\quad f^\sharp \quad} T$$

such that

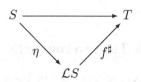

commutes where

$$\eta = \eta^\exists \qquad \eta = \eta^\forall$$

for the respective cases.

Proof   By Lemma 6.4.2 there is at most one such morphism $f^\sharp$. Thus it suffices to exhibit one such morphism. For each $X \in \mathcal{L}S$ let

$$f^\sharp(X) = \bigvee f[X] \qquad f^\sharp(X) = \bigwedge f[X']$$

for the respective cases. Almost trivially this *function* makes the triangle commute. We must show that $f^\sharp$ is a morphism of the appropriate kind. Thus we required

$$f^\sharp(\bigcup \mathcal{X}) = \bigvee f^\sharp[\mathcal{X}] \qquad f^\sharp(\bigcap \mathcal{X}) = \bigwedge f^\sharp[\mathcal{X}]$$

for each collection $\mathcal{X}$ of lower sections of S. These follow by straightforward calculations but a little more care is required for the $\forall$-case. $\square$

Exercises

6.4.1   Prove Lemma 6.4.1.

6.4.2   Prove Lemma 6.4.2.

6.4.3   Complete the proof of Theorem 6.4.3.

6.5 Three endo-functors on *Pos*

In the usual way Theorem 6.4.3 induces a pair of functors

$$\textbf{Pos} \longrightarrow \textit{Sup} \qquad\qquad \textbf{Pos} \longrightarrow \textit{Inf}$$

each of which is the left adjoint to the corresponding forgetful functor. In this section we look at the induced composite covariant endo-functors on *Pos*

$$\textbf{Pos} \longrightarrow \textit{Sup} \longrightarrow \textbf{Pos} \qquad\qquad \textbf{Pos} \longrightarrow \textit{Inf} \longrightarrow \textbf{Pos}$$

together with an associated contravariant endo-functor. Curiously each one has the same object assignment, namely

$$S \longmapsto \mathcal{L}S$$

the one that sends each poset to its poset of lower sections. This means we have to devise different names to distinguish between the three.

The contravariant functor is the easiest to describe, as follows.

6.5.1 Definition For each poset S let

$$\mathsf{I}S = \mathcal{L}S$$

be the set of lower sections of S. For each monotone map

$$S \xrightarrow{\quad f \quad} T$$

let

$$\begin{aligned} \mathcal{L}S &\xleftarrow{\quad \mathsf{I}(f) \quad} \mathcal{L}T \\ f^{\leftarrow}(Y) &\longleftarrow\!\!\!| \; Y \end{aligned}$$

be the induced inverse image function. □

It is a simple exercise to show that $\mathsf{I}(f)$ is monotone. (In fact, this is a particular case of the observation that each continuous map between topological spaces induces a monotone map between the topologies.) It is just as easy to show that I is an endo-functor on *Pos*.

Next we deal with the two covariant endo-functors on *Pos*.

6.5.2 Definition For each poset S let

$$\exists S = \mathcal{L}S \qquad \forall S = \mathcal{L}S$$

be the set of lower sections of S. For each monotone map

$$S \xrightarrow{\quad f \quad} T$$

let

$$LS \xrightarrow{\exists(f)} LT \qquad\qquad LS \xrightarrow{\forall(f)} LT$$
$$X \longmapsto \downarrow f[X] \qquad\qquad X \longmapsto (\uparrow f[X'])'$$

be the modified direct image function and the twisted direct image function, respectively. □

There are one or two things to be checked here. We first observe that

$$b \in \exists(f)(X) \iff (\exists x \in S)[b \le f(x) \,\&\, x \in X]$$
$$b \in \forall(f)(X) \iff (\forall x \in S)[f(x) \le b \Rightarrow x \in X]$$

for each $X \in LS$ and $b \in T$. Using these we can check that each of $\exists$ and $\forall$ is an endo-functor on **Pos**. We can also check that these are the functors induced by the completion functors

$$\textbf{Pos} \longrightarrow \textbf{Sup} \qquad\qquad \textbf{Pos} \longrightarrow \textbf{Inf}$$

respectively.

What is the point of this? There is more here than we first see.

6.5.3 Lemma   *For each monotone map*

$$S \xrightarrow{\ f\ } T$$

between posets, the three induced monotone maps

$$LS \xleftarrow{\ \ \ \exists(f)\ \ \ }_{\displaystyle \overset{}{\underset{\forall(f)}{\longleftarrow}}} LT$$

$$LS \xleftarrow{\;\;I(f)\;\;} LT$$

form a double poset adjunction.

Proof   We require

$$\exists(f)(X) \subseteq Y \iff X \subseteq I(f)(Y) \qquad I(f)(Y) \subseteq X \iff Y \subseteq \forall(f)(X)$$

for all $X \in LS$ and $Y \in LT$. □

Exercises

6.5.1   Verify the explicit quantifier descriptions of $\exists(f)$ and $\forall(f)$.
  Use these to show that each of $\exists(f)$ and $\forall(f)$ is monotone.
  Use these descriptions to show that $\exists$ and $\forall$ are endo-functors on **Pos**.

6.5.2   Complete the proof of Lemma 6.5.3.

6.6 Long strings of adjunctions

Lemma 6.5.3 shows that each monotone map

$$S \xrightarrow{\ f\ } T$$

between a pair of posets induces two adjunctions

$$\mathcal{L}S \xrightarrow{\ \exists(f) \dashv \mathsf{I}(f) \dashv \forall(f)\ } \mathcal{L}T$$

between the posets of lower sections. Thus we produce a monotone map $\mathsf{I}(f)$ which has both a left and a right adjoint. Furthermore, a simple example shows that these two adjoints need not be the same. Thus there are strings of adjunctions of length 2. This suggests a question. Can there be longer strings of adjunctions? In this section we see that there are, and there is one very important example of such strings.

Suppose we start with a poset adjunction.

$$S \underset{g}{\overset{f}{\rightleftarrows}} T$$

We may apply the construction of Lemma 6.5.3 to each of f, g in turn, and so produce two strings of adjunction of length 2.

$$\mathcal{L}S \overset{\exists(f)}{\underset{\forall(f)}{\leftarrow\!\!-\mathsf{I}(f)-\!\!\rightarrow}} \mathcal{L}T \qquad\qquad \mathcal{L}T \overset{\exists(g)}{\underset{\forall(g)}{\leftarrow\!\!-\mathsf{I}(g)-\!\!\rightarrow}} \mathcal{L}S$$

What is the connection between these two strings?

6.6.1 Lemma *For each poset adjunction $f \dashv g$, as above, we have*

$$\mathsf{I}(f) = \exists(g) \qquad \forall(f) = \mathsf{I}(g)$$

and the two induced 2-strings can be merged.

Each 1-string adjunction $f \dashv g$ produces a 3-string adjunction

$$\mathcal{L}S \overset{\exists(f)}{\underset{\forall(g)}{\overset{\leftarrow\!-\mathsf{I}(f)=\exists(g)-}{\underset{-\forall(f)=\mathsf{I}(g)\rightarrow}{\longrightarrow}}}} \mathcal{L}T$$

between the two posets of lower sections. We may repeat this as in Table 6.1.

Table 6.1 *A poset development*

$$
\begin{array}{c}
\text{—}\exists^4\text{→} \\[2pt]
\text{—}\exists^3\text{→} \\[2pt]
\text{—}\exists^2\text{→} \qquad \text{←}\exists^2 \mathsf{I}\text{—} \qquad \vdots \\[2pt]
\text{—}\exists\text{→} \qquad \text{←}\exists\mathsf{I}\text{—} \qquad \text{—}\exists\mathsf{I}^2\text{→} \\[2pt]
S \xrightarrow{\ \delta\ } \mathcal{L}S \xleftarrow{\ \mathsf{I}\ } \mathcal{L}^2 S \xrightarrow{\ \mathsf{I}^2\ } \mathcal{L}^3 S \xleftarrow{\ \mathsf{I}^3\ } \mathcal{L}^4 S \xleftarrow{\ \mathsf{I}^4\ } \\[2pt]
\text{—}\forall\text{→} \qquad \text{←}\mathsf{I}\forall\text{—} \qquad \text{—}\mathsf{I}^2\forall\text{→} \\[2pt]
\text{—}\forall^2\text{→} \qquad \text{←}\mathsf{I}\forall^2\text{—} \qquad \vdots \\[2pt]
\text{—}\forall^3\text{→} \\[2pt]
\text{—}\forall^4\text{→}
\end{array}
$$

Suppose we start with a 2-string $f \dashv g \dashv h$ of adjunctions. This induces a 5-string of adjunctions

$$\mathcal{L}S \ \text{—}\exists(f) \dashv \mathsf{I}(f) = \exists(g) \dashv \forall(f) = \mathsf{I}(g) = \exists(h) \dashv \forall(g) = \mathsf{I}(h) \dashv \forall(h) \to\ \mathcal{L}T$$

between the two posets of lower sections. This indicates that we can generate arbitrarily long strings of adjunctions between posets.

We start from any monotone map between posets

$$S \longrightarrow T$$

which we need not name. We hit this once with the construction of Lemma 6.6.1 to produce a 2-string between the first level completions.

$$\mathcal{L}S \quad \exists \dashv \mathsf{I} \dashv \forall \quad \mathcal{L}T$$

Here we haven't even written in the arrows. We now hit this with the construction to produce three 2-two strings between the second level completions, the posets of lower section of the poset of lower sections.

$$
\begin{array}{ccc}
 & \exists^2 \dashv \mathsf{I}\exists \dashv \forall\exists & \\
\mathcal{L}^2 S & \exists\mathsf{I} \dashv \mathsf{I}^2 \dashv \forall\mathsf{I} & \mathcal{L}^2 T \\
 & \exists\forall \dashv \mathsf{I}\forall \dashv \forall^2 &
\end{array}
$$

We know these merge to produce a 4-string of adjunctions.

$$\mathcal{L}^2 S \quad \exists^2 \dashv \exists\mathsf{I} \dashv \mathsf{I}^2 \dashv \mathsf{I}\forall \dashv \forall^2 \quad \mathcal{L}^2 T$$

Here we have made a choice in the way we name these arrows. We now hit this to produce a 6-string of adjunctions.

$$\mathcal{L}^2 S \quad \exists^3 \dashv \exists^2\mathsf{I} \dashv \exists\mathsf{I}^2 \dashv \mathsf{I}^3 \dashv \mathsf{I}^2\forall \dashv \mathsf{I}\forall^2 \dashv \forall^3 \quad \mathcal{L}^2 T$$

We hit this again to produce an 8-string, then a 10-string, and so on.

Table 6.2 *The simplicial category*

$$0 \xrightarrow{\ \delta_0^0\ } 1 \begin{array}{c} \xrightarrow{\ \delta_1^1\ } \\ \xrightarrow{\ \delta_0^1\ } \end{array} 2 \begin{array}{c} \xrightarrow{\ \delta_2^2\ } \\ \xrightarrow{\ \delta_1^2\ } \\ \xrightarrow{\ \delta_0^1\ } \end{array} 3 \begin{array}{c} \xrightarrow{\ \delta_3^3\ } \\ \xrightarrow{\ \delta_2^3\ } \\ \xrightarrow{\ \delta_1^3\ } \\ \xrightarrow{\ \delta_0^3\ } \end{array} 4 \quad \cdots$$

$$1 \xleftarrow{\ \sigma_0^1\ } 2 \begin{array}{c} \xleftarrow{\ \sigma_1^2\ } \\ \xleftarrow{\ \sigma_0^2\ } \end{array} 3 \begin{array}{c} \xleftarrow{\ \sigma_2^3\ } \\ \xleftarrow{\ \sigma_2^3\ } \\ \xleftarrow{\ \sigma_0^3\ } \end{array} 4 \quad \cdots$$

Of course, we haven't yet checked that there are no more equalities between the arrows beyond those we have indicated. To do that let's look at a particular case.

Suppose we start from

$$S \xrightarrow{\ \delta\ } \mathcal{L}S$$

an arbitrary poset S and an arbitrary monotone map δ to its poset of lower sections. By repeatedly hitting this we generate the development of Table 6.1. Let's consider a particular case of this. Let us take the empty poset for S. Then each of

$$\emptyset, \mathcal{L}\emptyset, \mathcal{L}^2\emptyset, \mathcal{L}^2\emptyset, \mathcal{L}^3\emptyset, \ldots$$

is linearly ordered with $0, 1, 2, 3, \ldots$ members. We can think of them as the natural numbers. There is only one possible map δ, the empty map. Thus we generate the collection of monotone maps as in Table 6.2. This is the simplicial category, an important gadget in algebraic geometry.

Exercises

6.6.1 Find an example of a monotone map f for which the two induced monotone maps $\exists(f), \forall(f)$ are not the same.

6.6.2 Prove Lemma 6.6.1.

6.6.3 Look up the standard construction of the simplicial category and check that it is the same as that in Table 6.2.

6.7 Two adjunctions for R-sets

In the second part of this chapter we look at the category $Set\text{-}R$ of R-sets and two adjunctions between it and the category Set of sets. These adjunctions convert each set into a free R-set and the cofree R-set, respectively. In this section we set up the basics and then get down to business in the next sections.

We have seen the notion of a monoid before, but there is no harm in repeating the definition.

6.7.1 Definition   A monoid is a structure

$$(R, \cdot, 1)$$

where R is a set furnished with an associative binary operation and a neutral element for the operation.   □

As is the custom, we usually hide the furnishings and speak of 'a monoid R'. We don't display the operation symbol in compounds, we write rs for $r \cdot s$ for elements $r, s \in R$. Using this convention we see that

$$(rs)t = r(st) \qquad 1r = r = r1$$

are the axioms for a monoid. Here, and below, we let r, s, t range over R. Because of the associativity we often leave out brackets and write

$$rst$$

for the two left hand compounds.

You can think of a monoid as a ring with the addition missing. Much of what we do here can be extended to rings, but sometimes that is quite a bit more complicated.

We now fix a monoid R.

6.7.2 Definition   A right R-set is a set A with a right R-action

$$A, R \longrightarrow A$$
$$a, r \longmapsto ar$$

an operation that combines an element $a \in A$ with an element $r \in R$ to return an element $ar \in A$. This action is required to satisfy

$$(as)r = a(sr) \qquad a1 = a$$

for each $a \in A$ and $r, s \in R$.   □

Notice that we are using concatenation for at least two different operations. If you find this confusing then for a while insert a different symbol for each different use of concatenation.

There is also a notion of a left R-set, where the action operates on the other side. We don't need those so here 'R-set' means 'right R-set'.

6.7.3 Definition Given two R-sets A and B, an R-morphism

$$A \xrightarrow{\quad l \quad} B$$

is a function l, as indicated, such that

$$l(ar) = l(a)r$$

for each $a \in A$ and $r \in R$.

This gives us the category

$$\textbf{\textit{Set-R}}$$

of R-sets and R-morphisms. Because here there is no danger of confusion we often say 'morphism' in place of 'R-morphism'. $\qquad\square$

We compare the category **Set-R** with the category

$$\textbf{\textit{Set}}$$

of sets and functions. Since each R-set A is a set furnished with some structure indexed by R, there is a forgetful functor

$$\textbf{\textit{Set}} \leftarrow U - \textbf{\textit{Set-R}}$$

which sends each R-set to its carrying set, and each R-morphism to its carrying function. In other words U forgets all the structure. In the following sections we show that U has a left adjoint and a right adjoint

$$\textbf{\textit{Set}} \overset{\overset{\textstyle \Sigma}{\longrightarrow}}{\underset{\underset{\textstyle \Pi}{\longrightarrow}}{\longleftarrow U \longrightarrow}} \textbf{\textit{Set-R}}$$

and so obtain two adjunctions

$$\Sigma \dashv U \dashv \Pi$$

which, for convenience, we refer to as the

upper   lower

adjunction, respectively.

We could quite quickly demonstrate the existence of these adjunctions but that is not the aim here. The purpose is to provide fairly simple illustrations of all the various aspects of adjunctions. We look at all the bits of gadgetry to see exactly what they do in these particular examples.

Many of the details are left as exercises.

Because the functor U is forgetful we often omit to name it. For instance, for an R-set A we sometimes write ΣA or ΠA for $(\Sigma \circ U)A$ or $(\Pi \circ U)A$, respectively. If you find this confusing then insert U in the places where you think it should be.

Several kinds of algebras, a set furnished with some operations, are R-sets for a particular monoid R. We have met one of these before, and Exercise 6.7.1 gives three more. You may find it helpful to sort out the details for these particular monoids as you read the following sections.

Exercises

6.7.1   Consider the following four kinds of algebras.

(a) An involution algebra is a structure

$$(A, (\cdot)^{\bullet}) \qquad a^{\bullet\bullet} = a$$

as on the left where the identity on the right holds for all $a \in A$.

(b) An idempotent algebra is a structure

$$(A, (\cdot)^{\bullet}) \qquad a^{\bullet\bullet} = a^{\bullet}$$

as on the left where the identity on the right holds for all $a \in A$.

(c) A 2-step involution algebra is a structure

$$(A, (\cdot)^{\bullet}, {}^{\bullet}(\cdot)) \qquad a^{\bullet\bullet} = {}^{\bullet}a \quad {}^{\bullet}(a^{\bullet}) = ({}^{\bullet}a)^{\bullet} \quad {}^{\bullet\bullet}a = a$$

as on the left where the identities on the right hold for all $a \in A$.

(d) A 2-step idempotent algebra is a structure

$$(A, (\cdot)^{\bullet}, {}^{\bullet}(\cdot)) \qquad a^{\bullet\bullet} = {}^{\bullet}a \quad {}^{\bullet\bullet}a = a^{\bullet} \quad {}^{\bullet}(a^{\bullet}) = ({}^{\bullet}a)^{\bullet} = a$$

as on the left where the identities on the right hold for all $a \in A$.

Show that each of these algebras is an R-set for a particular monoid R, which you should describe in detail.

Produce another examples of this kind of algebra.

6.8 The upper left adjoint

In this section we produce the functor Σ but, as yet, we don't show it is left adjoint to U. We also produce assignments

$$X \xrightarrow{\ \eta X\ } (U \circ \Sigma)X \qquad (\Sigma \circ U) \xrightarrow{\ \epsilon_A\ } A$$

for each set X and each R-set A. In due course we see that these are the unit and the counit of the upper adjunction $\Sigma \dashv U$. We don't attempt a direct analysis, we wander around gathering all the relevant properties.

6.8.1 Definition For each set X let

$$\Sigma X = X \times R$$

the set of ordered pairs (x, r) for $x \in X, r \in R$. Also let

$$\Sigma X, R \longrightarrow \Sigma X$$
$$(x, r), s \longmapsto (x, rs)$$

for each $x \in X$ and $r, s \in R$. $\qquad\qquad\square$

It is easy to check that ΣX with this action is an R-set. In due course we will show that ΣX is the free R-set generated by X. On general grounds it then follows that Σ is the object assignment of a functor from *Set* to *Set-R*. Here we don't rely on that. We produce the arrow assignment directly.

6.8.2 Definition For each function

$$Y \xrightarrow{\ g\ } X$$

between sets let

$$\Sigma Y \xrightarrow{\ \Sigma(g)\ } \Sigma X$$

be given by

$$\Sigma(g)(y, r) = (g(y), r)$$

for each $y \in Y$ and $r \in R$. $\qquad\qquad\square$

We need to check that $\Sigma(g)$ is an R-morphism, and the two assignment combine to form a functor. These details are left as exercises.

We now produce assignments

$$X \xrightarrow{\ \eta X\ } (U \circ \Sigma)X \qquad (\Sigma \circ U) \xrightarrow{\ \epsilon_A\ } A$$

for each set X and each R-set A. These turn out to be the unit and the counit of the upper adjunction $\Sigma \dashv U$.

We begin by looking at η.

6.8.3 Definition　For each set X let η_X as on the left

$$X \xrightarrow{\ \eta_X\ } (U \circ \Sigma)X \qquad \eta_X(x) = (x, 1)$$

be the function given by the assignment on the right for each $x \in X$.　　　□

This is just a function, it need not have any morphism properties.

Consider any $(x, r) \in \Sigma X$. Remembering the action on Σ we have

$$(x, r) = (x, 1r) = (x, 1)r = \eta_X(x)r$$

to show how η_X picks out a generating set of ΣX.

6.8.4 Lemma　*For each set X the assignment η_X is 'sufficiently epic'. That is,*

$$f \circ \eta_X = g \circ \eta_X \implies f = g$$

for each parallel pair

$$\Sigma X \; \underset{g}{\overset{f}{\rightrightarrows}} \; A$$

of R-morphisms.

The following shows that ΣX is the free R-set generated by X via η_X.

6.8.5 Theorem　*For each function*

$$X \xrightarrow{\ g\ } A$$

from a set X to an R-set A, there is a unique R-morphism

$$\Sigma X \xrightarrow{\ g^\sharp\ } A$$

such that

$$
\begin{array}{ccc}
X & \xrightarrow{\ g\ } & A \\
& {\scriptstyle \eta_X}\searrow & \nearrow{\scriptstyle g^\sharp} \\
& \Sigma X &
\end{array}
$$

commutes in **Set**.

As is usual in this kind of situation, we have omitted the underlying functor U. If you find this confusing, then simply insert U at the appropriate places

How can we prove Theorem 6.8.5? By Lemma 6.8.4 there is at most one such morphism $g^\sharp$, so it suffices to exhibit one. Consider any element

$$(x, r) = \eta_X(x)r$$

of ΣX. Remembering that $g^\sharp$ must be an R-morphism we have

$$g^\sharp(x, r) = g^\sharp(\eta_X(x)r) = g^\sharp(\eta_X(x))r = (g^\sharp \circ \eta_X)(x)r = g(x)r$$

which shows us the only possible function that will work.

On general grounds Theorem 6.8.5 ensures that the object assignment Σ of Definition 6.8.1 fills out to a functor

$$\textit{Set} \xrightarrow{\ \Sigma\ } \textit{Set-R}$$

for which

$$Id_{\textit{Set}} \xrightarrow{\ \eta_\bullet\ } (U \circ \Sigma)$$

is natural. We can check that this functor Σ is the one we first thought of, and show by direct calculation that η is natural. For this second part we need to check that each function g induces a commuting square

$$
\begin{array}{ccc}
Y & \xrightarrow{\ \eta_Y\ } & \Sigma Y \\
{\scriptstyle g}\downarrow & & \downarrow{\scriptstyle \Sigma(g)} \\
X & \xrightarrow[\ \eta_X\ ]{} & \Sigma X
\end{array}
$$

in *Set*. This is more or less trivial.

Let us now look at the counit. In some ways this is more interesting.

6.8.6 Definition For each R-set A let ϵ_A, as on the left,

$$(\Sigma \circ U)A \xrightarrow{\ \epsilon_A\ } A \qquad\qquad \epsilon_A(a, r) = ar$$

be the function given as on the right for each $a \in A$ and $r \in R$. □

This construction starts with an R-set A. We forget its structure to produce a set UA and then we furnish that as an R-set using Σ. Thus we obtain an R-set of pairs (a, r) for $a \in A$ and $r \in R$. This shows that the construction of ϵ_A does make sense, as a function. However, we need to check that ϵ_A is an R-morphism. In other words, we require

$$\epsilon_A\big((a, r)s\big) = \big(\epsilon_A(a, r)\big)s$$

for all $a \in A$ and $r, s \in R$. This follows by unravelling the definitions.

We need to check that $\epsilon_\bullet$ is natural.

$$(\Sigma \circ U) \xrightarrow{\;\epsilon_\bullet\;} Id_{Set\text{-}R}$$

In other words, we need to check that for each R-morphism f the square

$$
\begin{array}{ccc}
\Sigma A & \xrightarrow{\;\epsilon_A\;} & A \\
{\scriptstyle \Sigma(f)}\downarrow & & \downarrow{\scriptstyle f} \\
\Sigma B & \xrightarrow[\;\epsilon_B\;]{} & B
\end{array}
$$

commutes. A proof of this is not exactly strenuous.

Exercises

6.8.1   Show that Definition 6.8.1 does produce an R-set ΣX.

6.8.2   Show Definitions 6.8.1 and 6.8.2 give a functor from *Set* to *Set-R*.

6.8.3   Prove Lemma 6.8.4.

6.8.4   Prove Theorem 6.8.5.

6.8.5   Show that the arrow assignment of the functor Σ ensured by Theorem 6.8.5 is that given by Definition 6.8.2.

6.8.6   Prove directly that the unit η is natural.

6.8.7   Show that the construction of Definition 6.8.6 does produce an R-morphism.

6.8.8   Prove directly that the counit ϵ is natural.

6.8.9   Show that for each R-morphism

$$\Sigma X \xrightarrow{\;f\;} A$$

from a free R-set to an arbitrary R-set, there is a unique function

$$X \xrightarrow{\;f_\flat\;} U A$$

such that

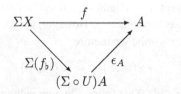

commutes in *Set-R*.

6.9 The upper adjunction

We have a pair of functors

$$\mathbf{Set} \xrightleftharpoons[U]{\Sigma} \mathbf{Set}\text{-}R$$

and our aim is to show that these form an adjoint pair. We have produced natural transformations

$$X \xrightarrow{\eta_X} (U \circ \Sigma)X \qquad (\Sigma \circ U)A \xrightarrow{\epsilon_A} A$$

which we hope will be the unit and counit of the adjunction. In fact, Theorem 6.8.5 or Exercise 6.8.9 is enough to ensure we do have such an adjunction. In this section we show that we don't need either of those results. We use Theorem 5.4.5. Thus we show

$$\epsilon_{\Sigma X} \circ \Sigma(\eta_X) = id_{\Sigma X} \qquad U(\epsilon_A) \circ \eta_{UA} = id_{UA}$$

for each set X and R-set A. These are not difficult to verify, but the left hand identity does require a bit more care. Let's have a look at that.

We need to understand the composite

$$\Sigma X \xrightarrow{\Sigma(\eta_X)} (\Sigma \circ U \circ \Sigma)X \xrightarrow{\epsilon_{\Sigma X}} \Sigma X$$

for an arbitrary set X. Now ΣX is a set of ordered pairs (x, r) for $x \in X$ and $r \in R$. This means that

$$(\Sigma \circ U \circ \Sigma)X$$

is a set of ordered pairs (l, r) where l is already an ordered pair. Starting with $(x, r) \in \Sigma X$ we use $\Sigma(\eta_X)$ to produce such a pair (l, r) in the central component, and then hit this with $\epsilon_{\Sigma X}$ to produce an ordered pair in ΣX. We need to check that this is the starting ordered pair.

In the remainder of this section we look at the upper transitions to obtain what is usually taken as the 'official' definition of an adjunction. We set up an inverse pair of transformations

$$\mathbf{Set}[X, UA] \xrightleftharpoons[(\cdot)_\flat]{(\cdot)^\natural} \mathbf{Set}\text{-}R[\Sigma X, A]$$

for each set X and R-set A, and show these are natural for variations of the objects. We also check directly various other properties associated with such transitions.

At the function level what might the two transitions be?

$$g \longmapsto g^\sharp$$
$$(X \to A) \qquad (X \times R \to A)$$
$$f_\flat \longleftarrow \hspace{1em}\dashv f$$

The simplest suggestion seems to be

$$g^\sharp(x,r) = g(x)r \qquad f_\flat(x) = f(x,1)$$

for each $x \in X$ and $r \in R$. We are going to check directly that these are the correct functions. But before we do that let's use what we have already done to give a quick proof of correctness. By Lemma 5.4.3 we know that we must have

$$g^\sharp = \epsilon_A \circ \Sigma(g) \qquad f_\flat = U(f) \circ \eta_X$$

for each function g and morphism f, as above. By unravelling these compounds we see that the suggestions above are correct.

We now go through the details of the direct verifications.

Consider any function

$$g : X \longrightarrow A$$

as above. Thus X is an arbitrary set but A is (the carrier of) an R-set. The suggested construction certainly gives a function

$$g^\sharp : \Sigma X \longrightarrow A$$

but we require this to be an R-morphism, that is

$$g^\sharp\big((x,r)s\big) = g^\sharp(x,r)s$$

for each $x \in X$ and $r, s \in R$. This follows by a simple calculation.

Next consider any morphism

$$\Sigma X \xrightarrow{\quad f \quad} A$$

as above. The construction does give a function

$$f_\flat : X \longrightarrow A$$

as we want.

We require these two transitions to form an inverse pair, that is

$$g^\sharp{}_\flat = g \qquad f_\flat{}^\sharp = f$$

for each function g and morphism f, as above. These follow by simple calculations, but the right hand one does have just a little more content.

Finally, we check the naturality of these two transition assignments. Of course since each is the inverse of the other we need only check that one is natural, but it is instructive to do both calculations.

We check the two conditions ($\sharp$) and ($\flat$) of Tables 5.1 and 5.2 on pages 165 and 167. Those tables uses a slightly different notation, so let's start again.

For ($\sharp$) consider the square

$$
\begin{array}{cccc}
X & Set[X,UA] \xrightarrow{\ \ (\cdot)^{\sharp}\ \ } Set\text{-}R[\Sigma X, A] & A \\
\Big\uparrow k & U(l)\circ - \circ k \Big\downarrow \qquad \Big\downarrow l\circ - \circ \Sigma(k) & \Big\downarrow l \\
Y & Set[Y,UB] \xrightarrow[\ \ (\cdot)^{\sharp}\ \ ]{} Set\text{-}R[\Sigma Y, B] & B
\end{array}
$$

induced by a function k and a morphism l, as indicated. We must show that this square commutes, that is

$$\big(U(l)\circ g\circ k\big)^{\sharp} = \big(l\circ g^{\sharp}\circ \Sigma(k)\big)$$

for each function

$$X \xrightarrow{\ \ g\ \ } A$$

from the top left hand corner. To check this we evaluate both sides at an arbitrary member (y,r) of ΣY, and remember that l is a morphism.

For ($\flat$) consider the square

$$
\begin{array}{cccc}
X & Set[X,UA] \xleftarrow{\ \ (\cdot)_{\flat}\ \ } Set\text{-}R[\Sigma X, A] & A \\
\Big\uparrow k & U(l)\circ - \circ k \Big\downarrow \qquad \Big\downarrow l\circ - \circ \Sigma(k) & \Big\downarrow l \\
Y & Set[Y,UB] \xleftarrow[\ \ (\cdot)_{\flat}\ \ ]{} Set\text{-}R[\Sigma Y, B] & B
\end{array}
$$

induced by a function k and a morphism l, as indicated. We must show that this square commutes, that is

$$\big(l\circ f_{\flat}\circ k\big) = \big(l\circ f\circ \Sigma(k)\big)_{\flat}$$

for each morphism

$$\Sigma X \xrightarrow{\ \ f\ \ } A$$

from the top right hand corner. To check this we evaluate both sides at an arbitrary member y of Y.

This completes the discussion of

$$\Sigma \dashv U$$

the upper adjunction.

Exercises

6.9.1 Verify the two identities involving η and ϵ that ensure we do have an adjunction.

6.9.2 For the suggested transitions show that

$$g^{\sharp} = \epsilon_A \circ \Sigma(g) \qquad f_{\flat} = U(f) \circ \eta_X$$

do hold for each function g and morphism f, as above.

6.9.3 Show that for each function g, as above, the suggested function $g^{\sharp}$ is a morphism.

6.9.4 Show that $(\cdot)^{\sharp}$ and $(\cdot)_{\flat}$ form an inverse pair of transitions.

6.9.5 Show that each of the transitions $(\cdot)^{\sharp}$ and $(\cdot)_{\flat}$ is natural.

6.10 The lower right adjoint

We now turn to the lower adjunction.

$$\Pi \dashv U$$

Remember that the functors seem to go the wrong way, so the re-drawn picture

$$\textit{Set-R} \xrightarrow{\ \ U\ \ } \textit{Set}$$
$$\xleftarrow[\ \ \Pi\ \ ]{}$$

might help us to avoid a bit of confusion. In this section we produce the functor Π, but we don't start to deal with the adjunction properties. We also produce assignments

$$A \xrightarrow{\ \eta_A\ } (\Pi \circ U)A \qquad (U \circ \Pi)X \xrightarrow{\ \epsilon_X\ } X$$

for each R-set A and set X. Eventually we see that these are the unit and the counit of the lower adjunction $\Pi \dashv U$. However, as with the upper adjunction we first wander around their properties getting to know them.

The construction Π must convert an arbitrary set X into an R-set ΠX. To do that we first need the carrier.

6.10.1 Definition For each set X let

$$\Pi X = \mathbf{Set}[UR, X]$$

the set of all functions

$$h : R \longrightarrow X$$

from the *set* R to X. □

We need to furnish ΠX as an R-set. At first the construction and notation we use might look odd, but you will soon see why we use it.

6.10.2 Definition For each set X, each function

$$h : R \longrightarrow X$$

and each $r \in R$, let

$$h^r : R \longrightarrow X$$

be the function given by

$$h^r(s) = h(rs)$$

for each $s \in R$. □

This certainly produces an assignment

$$\Pi X, R \longrightarrow \Pi X$$
$$h, r \longmapsto h^r$$

and a couple of simple calculations shows that it is an R-action. This is the object assignment of the functor. Here is the arrow assignment.

6.10.3 Definition For each function

$$X \xrightarrow{\;g\;} Y$$

let

$$\Pi X \xrightarrow{\;\Pi(g)\;} \Pi Y$$

be the function given by composition, that is

$$\Pi(g)(h) = g \circ h$$

for each $h \in \Pi X$. □

Of course, we need $\Pi(g)$ to be an R-morphism, that is

$$\Pi(g)(h^r) = \big(\Pi(g)(h)\big)^r$$

for each $h \in \Pi X$ and $r \in R$. This follows by evaluating both sides at an arbitrary $s \in R$.

We need to know that the two assignments Π combine to form a functor. In fact, at this stage there is nothing we have to prove. Observe that at the *Set* level Π is just the hom-functor

$$\textbf{\textit{Set}}[UR, -]$$

induced by the set R. We have checked that both the object assignment and the arrow assignment do the right thing, so we have what we want. This is an example of an enriched hom-functor.

We now produce assignments

$$A \xrightarrow{\eta_A} (\Pi \circ U)A \qquad (U \circ \Pi)X \xrightarrow{\epsilon_X} X$$

for each R-set A and set X. Eventually we see that these are the unit and the counit of the lower adjunction $\Pi \dashv U$.

What might these two assignments be? Once we think about it we see that there is not much choice. Consider $a \in A$. We require a function

$$\eta_A(a) : R \longrightarrow A$$

which can be evaluated at an arbitrary $r \in R$. For ϵ_X consider any typical $h \in (U \circ \Pi)X$, an arbitrary function

$$h : R \longrightarrow X$$

from which we must obtain an element of X. The choices are obvious.

6.10.4 Definition   For each R-set A and set X let

$$A \xrightarrow{\eta_A} (\Pi \circ U)A \qquad (U \circ \Pi)X \xrightarrow{\epsilon_X} X$$

be the functions given by

$$\eta_A(a)(r) = ar \qquad \epsilon_X(h) = h(1)$$

for each $a \in A, r \in R$, and $h \in \Pi X$.   □

We want η_A to be an R-morphism. Remembering how $(\Pi \circ U)A$ is structured this requirement is

$$\eta_A(ar) = \eta_A(a)^r$$

for all $a \in A$ and $r \in R$. This can be checked by evaluating both sides at an arbitrary member of R.

There are some properties that η and ϵ must have. We look at η first.
We require the family $\eta_\bullet$ of morphisms to be natural. Thus the square

$$
\begin{array}{ccc}
A & \xrightarrow{\ \eta_A\ } & \Pi A \\
{\scriptstyle f}\big\downarrow & & \big\downarrow{\scriptstyle \Pi(f)} \\
B & \xrightarrow[\ \eta_B\ ]{} & \Pi B
\end{array}
$$

must commute for each morphism f. This requirement is

$$\eta_B \circ f = \Pi(f) \circ \eta_A$$

in equational form. To check this we evaluate both sides at an arbitrary $a \in A$
to produce a pair of functions

$$R \longrightarrow B$$

which must be equal. So we evaluate each of these at an arbitrary $r \in R$.

In the usual way the existence of a unit with an appropriate universal property ensures we do have an adjunction.

6.10.5 Theorem   *For each R-set A, set X and morphism*

$$A \xrightarrow{\ f\ } \Pi X$$

there is a unique function

$$U A \xrightarrow{\ f^\sharp\ } X$$

such that

$$
\begin{array}{ccc}
A & \xrightarrow{\quad f \quad} & \Pi X \\
 & {\scriptstyle \eta_A}\searrow \quad \nearrow {\scriptstyle \Pi(f^\sharp)} & \\
 & (\Pi \circ U)A &
\end{array}
$$

commutes in **Set**-*R*.

Proof   Let's first show that there is at most one such function $f^\sharp$.
Suppose there is some such $f^\sharp$. Then

$$\Pi(f^\sharp) \circ \eta_A = f$$

and hence

$$\Pi(f^\sharp)(\eta_A(a)) = f(a)$$

for each $a \in A$. This gives

$$f^{\sharp} \circ \eta_A(a) = f(a)$$

by the construction of Π. Each of the two sides of this equality is a function $R \longrightarrow A$, and hence we may evaluate at 1. Remembering the construction of η this gives

$$f^{\sharp}(a) = f^{\sharp}(\eta_A(a)(1)) = f(a)(1)$$

to determine $f^{\sharp}$ uniquely.

We use this as the definition of the function $f^{\sharp}$, so all that remains is to show that the induced triangle does commute.   □

We now check the various properties of ϵ_X. Since U is the forgetful functor we see that

$$\Pi X \xrightarrow{\ \epsilon_X\ } X$$

is just a function, so we do not require any morphisms properties. We do require that the whole family $\epsilon_{\bullet}$ is natural. The induced square

$$
\begin{array}{ccc}
\Pi X & \xrightarrow{\ \epsilon_X\ } & X \\
{\scriptstyle \Pi(g)} \downarrow & & \downarrow {\scriptstyle g} \\
\Pi Y & \xrightarrow[\ \epsilon_Y\ ]{} & Y
\end{array}
$$

must commute, that is

$$\epsilon_Y \circ \Pi(g) = g \circ \epsilon_X$$

in equational form. To check this we evaluate both sides at an arbitrary member $h : R \longrightarrow X$ of the top left hand corner of the square.

These functions $\epsilon_{\bullet}$ are part of a cofree construction. Here are some of the details of that.

6.10.6 Lemma *For each set X the assignment ϵ_X is 'sufficiently monic' in the sense that*

$$\epsilon_X \circ k = \epsilon_X \circ l \Longrightarrow k = l$$

for each parallel pair

$$
A \underset{l}{\overset{k}{\rightrightarrows}} \Pi X
$$

of R-morphisms.

Proof   Assuming

$$\epsilon_X \circ k = \epsilon_X \circ l$$

we have

$$\epsilon_X\big(k(a)\big) = \epsilon_X\big(l(a)\big)$$

that is

$$k(a)(1) = l(a)(1)$$

for each $a \in A$. Each of k, l is a 2-step function

$$A \longrightarrow R \longrightarrow X$$

so we require

$$k(a)(r) = l(a)(r)$$

for each $a \in A$ and $r \in R$. By remembering how ΠX is structured we see how the above identity leads to this more general identity.   □

A few more calculations gives the following factorization result.

6.10.7 Theorem   *For each function*

$$A \xrightarrow{\;\;g\;\;} X$$

from an R-set A to a set X, there is a unique morphism

$$A \xrightarrow{\;\;g_\flat\;\;} \Pi X$$

such that

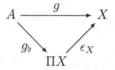

commutes in **Set**.

Proof   By Lemma 6.10.6 there is at most one such morphism $g_\flat$.

Whatever it is $\flat$ is a 2-step function

$$A \longrightarrow R \longrightarrow X$$

which must be evaluated first at an arbitrary $a \in A$ and then at an arbitrary $r \in R$ to return a value in X. There is an obvious choice for such a function. A few calculations shows that this choice does work.   □

As set up in this section the families $\eta_\bullet$ and $\epsilon_\bullet$ don't seem to be doing anything special. Of course, later we will see that they are the unit and counit of an adjunction.

Exercises

6.10.1   Show that for each set X the construction of ΠX does produce an R-set.

6.10.2   Show that the construction of Definition 6.10.3 really does produce an R-morphism.

6.10.3   Show that the construction of Definition 6.10.4 does produce a morphism η_A.

6.10.4   Show that the family $\eta_\bullet$ of morphisms is natural.

6.10.5   Complete the proof of Theorem 6.10.5.

6.10.6   Show that the family $\epsilon_\bullet$ of morphisms is natural.

6.10.7   Complete the proof of Lemma 6.10.6.

6.10.8   Complete the proof of Theorem 6.10.7.

6.11 The lower adjunction

We have a pair of functors

$$Set\text{-}R \underset{\Pi}{\overset{U}{\rightleftarrows}} Set$$

and our aim is to show that these form an adjoint pair. We have produced natural transformations

$$A \xrightarrow{\ \eta_a\ } (\Pi \circ U)A \qquad (U \circ \Pi)X \xrightarrow{\ \epsilon_x\ } X$$

which we will show are the unit and counit of the adjunction. In fact, Theorems 6.10.5 and 6.10.7 are enough to do this. However, in this section we don't take that route. We use Theorem 5.4.5. Thus we show

$$\epsilon_{\Pi A} \circ U(\eta_A) = id_{UA} \qquad \Pi(\epsilon_x) \circ \eta_{\Pi X} = id_{\Pi X}$$

for each R-set A and set X.

The left hand equality follows by evaluating each side at an arbitrary $a \in A$. The calculations are straightforward. The right hand equality follows by a similar method but does require a little more care.

We have a composite

$$\Pi X \xrightarrow{\eta_{\Pi X}} (\Pi \circ U \circ \Pi)X \xrightarrow{\Pi(\epsilon_X)} \Pi X$$

for an arbitrary set X. We evaluate this at an arbitrary member

$$h : R \longrightarrow X$$

of ΠX to produce another member of ΠX. We evaluate this at an arbitrary member $r \in R$ and hope to show that the result is $h(r)$.

In the remainder of this section we look at the 'official' definition of an adjunction. We require an inverse pair of transitions

$$\textbf{\textit{Set}-}R[A, \Pi X] \underset{(\cdot)_\flat}{\overset{(\cdot)^\natural}{\rightleftarrows}} \textbf{\textit{Set}}[UA, X]$$

for each R-set A and set X, and these must be natural for variations of the objects. We also check directly various other properties associated with such transitions. In fact, we already know what these assignments must be, since they must fit into Theorems 6.10.5 and 6.10.7. Thus we will show that the following two constructions do the job.

6.11.1 Definition For each R-set A and set X, and each

morphism function

$$A \xrightarrow{f} \Pi X \qquad A \xrightarrow{g} X$$

we set

$$f^\natural(a) = f(a)(1) \qquad g_\flat(a)(r) = g(ar)$$

for each $a \in A$ and $r \in R$. □

At the *Set*-level we do have

$$f^\natural : A \longrightarrow X \qquad g_\flat : A \longrightarrow R \longrightarrow X$$

and our job now is to check all the other requirements.

By Lemma 5.4.3 if Definition 6.11.1 is correct, then we must have

$$f^\natural = \epsilon_x \circ U(f) \qquad g_\flat = \Pi(g) \circ \eta_A$$

for each morphism f and function g, as above. By unravelling these compounds we see that the suggestions above are correct.

We now go through the various other details of the direct verifications.

At some point we have to show that these two transition form an inverse pair, that is

$$f^\natural{}_\flat = f \qquad g_\flat{}^\natural = g$$

for each morphism f and function g. These are not difficult to verify, but one of them does require a little bit of thought.

The next thing we should check is that $(\cdot)_\flat$ does return a morphism. To do this we remember how a cofree R-set ΠX is structured.

Finally, we check the naturality of these two transition assignments. As usual, since each is the inverse of the other we need only check that one is natural, but it is instructive to do both calculations. We check the conditions $(\sharp)$ and $(\flat)$ of Tables 5.1 and 5.2 on pages 165 and 167.

For $(\sharp)$ consider the square

$$
\begin{array}{cccc}
A & \textbf{\textit{Set-}}R[A,\Pi X] \xrightarrow{\ \ (\cdot)^\sharp\ \ } \textbf{\textit{Set}}[UA,X] & X \\[2mm]
k\Big\uparrow & \ \Pi(l)\circ - \circ k\Big\downarrow \qquad\qquad l\circ - \circ U(k)\Big\downarrow & \Big\downarrow l \\[2mm]
B & \textbf{\textit{Set-}}R[B,\Pi Y] \xrightarrow[\ \ (\cdot)^\sharp\ \ ]{} \textbf{\textit{Set}}[UB,Y] & Y
\end{array}
$$

induced by a morphism k and a function l, as indicated. We must show that this square commutes, that is

$$
\bigl(\Pi(l)\circ f\circ k\bigr)^\sharp = \bigl(l\circ f^\sharp\circ U(k)\bigr)
$$

for each morphism

$$
A \xrightarrow{\ \ f\ \ } \Pi X
$$

from the top left hand corner. To check this we evaluate both sides at an arbitrary member $b \in B$.

For $(\flat)$ consider the square

$$
\begin{array}{cccc}
A & \textbf{\textit{Set-}}R[A,\Pi X] \xleftarrow{\ \ (\cdot)_\flat\ \ } \textbf{\textit{Set}}[UA,X] & X \\[2mm]
k\Big\uparrow & \ \Pi(l)\circ - \circ k\Big\downarrow \qquad\qquad l\circ - \circ U(k)\Big\downarrow & \Big\downarrow l \\[2mm]
B & \textbf{\textit{Set-}}R[B,\Pi Y] \xleftarrow[\ \ (\cdot)_\flat\ \ ]{} \textbf{\textit{Set}}[UB,Y] & Y
\end{array}
$$

induced by a morphism k and a function l, as indicated. We must show that this square commutes, that is

$$
\Pi(l)\circ g_\flat\circ k = \bigl(l\circ g\circ U(k)\bigr)_\flat
$$

for each function

$$
A \xrightarrow{\ \ g\ \ } \Pi X
$$

from the top right hand corner. To check this we evaluate both sides first at an arbitrary $b \in B$ and then an arbitrary $r \in R$.

This completes the discussion of

$$U \dashv \Pi$$

the lower adjunction.

Exercises

6.11.1 Verify the two identities involving ϵ and η that ensure we do have an adjunction.

6.11.2 For the suggested transitions show that

$$f^{\sharp} = \epsilon_x \circ U(f) \qquad g_{\flat} = \Pi(g) \circ \eta_A$$

do hold for each morphism f and function g, as above.

6.11.3 Show that $(\cdot)^{\sharp}$ and $(\cdot)_{\flat}$ form an inverse pair of transitions.

6.11.4 Show that for each function g, as above, the suggested function $g_{\flat}$ is a morphism.

6.11.5 Show that each of the transitions $(\cdot)^{\sharp}$ and $(\cdot)_{\flat}$ is natural.

6.12 Some final projects

In this final section I will suggest various problems you might want to look at. In each case the solution is known, but you will learn something from your investigations. For most of the problems there is something extra that has to be done.

There are various refinements of the constructions of Sections 6.1 to 6.4. For instance, suppose we do not need to complete a poset fully, but require only suprema or infima for finite subsets. This can be obtained by modifying the two constructions, but there is something extra that has to be done. Such a completion may not be achievable in one step. For another variant, suppose we want to produce suprema for directed sets. What do we do?

There is another aspect of these completions that we should not ignore. Suppose we want to complete a poset in some sense and that poset already has some suprema or infima. Suppose we wish to preserve some of these in the constructed poset. What do we do? In the literature you will find a construction, the MacNeille completion, which completes a poset and preserves any suprema that the given poset may have. That is not the answer, for that construction is *not* functorial.

There are two extensions of the constructions of Sections 6.7 to 6.11.
For the first of these consider an arbitrary monoid morphism

$$S \xrightarrow{\phi} R$$

between a pair of monoids. Associated with the two monoids we have a pair of
categories

$$\textit{Set-S} \qquad \textit{Set-R}$$

the S-sets and the R-sets respectively. Observe that if S is the singleton monoid
then ϕ is uniquely determined and *Set-S* is just *Set*.

The morphism ϕ induces a functor

$$\textit{Set-S} \xleftarrow{\quad \Phi \quad} \textit{Set-R}$$

the restriction of scalars which we have met before. The construction of Φ is
part of Exercise 3.3.14. This functor has a left and a right adjoint

$$\textit{Set-S} \xleftarrow{\quad \Phi \quad} \textit{Set-R}$$
$$\Sigma \nearrow \qquad \Pi \searrow$$

which you should sort out. When S is the singleton monoid this reduces to the
double adjunction of Sections 6.7 to 6.11.

This can be generalized even further. Consider a ring morphism

$$S \xrightarrow{\phi} R$$

between a pair of rings. We have the two associated categories

$$\textit{Mod-S} \qquad \textit{Mod-R}$$

of modules over S and R, respectively. The morphism ϕ induces a functor

$$\textit{Mod-S} \xleftarrow{\quad \Phi \quad} \textit{Mod-R}$$

also called restriction of scalars. This has a left and a right adjoint

$$\textit{Mod-S} \xleftarrow{\quad \Phi \quad} \textit{Mod-R}$$
$$\Sigma \qquad \Pi$$

where Σ is given by a tensor product and Π is an enriched hom-functor. There
are several parts of mathematics, both algebraic and geometric, where these
functors are needed.

Bibliography

J. Adamek, H. Herrlich, and G. E. Strecker (2004): *Abstract and Concrete Categories: The Joy of Cats*, published online.

S. Awodey (2010): *Category Theory* (Second Edition), Oxford University Press.

M. Barr and C. Wells (1985): *Toposes, Triples and Theories*, Springer.

M. Barr and C. Wells (1990): *Category Theory for Computing Science*, Prentice Hall.

A. J. Berrick and M. E. Keating (2000): *Categories and Modules*, Cambridge University Press.

F. Borceux (1994): *Handbook of Categorical Algebra*, three volumes, Cambridge University Press.

S. Eilenberg and S. MacLane (1945): General theory of natural equivalences, *Transactions of the American Mathematical Society*, **58**, 231–294.

F. William Lawvere and Stephen H. Schanuel (1997): *Conceptual Mathematics, A First Introduction to Categories*, Cambridge University Press.

S. Mac Lane (1998): *Categories for the Working Mathematician* (Second Edition), Springer.

C. McClarty (1995): *Elementary Categories, Elementary Toposes*, Oxford University Press.

B. Mitchell (1965): *Theory of Categories*, Academic Press.

B. Pareigis (1970): *Categories and Functors*, Academic Press.

Index

Notation

C an arbitrary category, 1

C^{op} opposite of category C, 32

$\widehat{C}$ presheaves on a category, 96

$C^{\downarrow}$ arrow category from C, 19

C^{∇} category of ∇-diagrams from C, 115

$(C \downarrow S)$ slice category over S, 20

$(S \downarrow C)$ slice category under S, 20

$(U \downarrow L)$ comma category, 86

$C[A, B]$ hom-set from A to B
  in category C, 4

$[A, B]$ hom-set from A to B, 4

$\mathrm{Hom}_C[A, B]$ hom-set from A to B
  in category C, 5

Arw the arrows of a category, 1

Obj the objects of a category, 1

$source$ of an arrow, 2

$target$ of an arrow, 2

$- \circ -$ composition of arrows, 3

id_A identity arrow on A, 3

$\boldsymbol{id}_A$ identity arrow on A, 3

1_A identity arrow on A, 3

$\boldsymbol{Src}$ source category of a functor, 73

$\boldsymbol{Trg}$ target category of a functor, 73

Δ diagonal functor, 78

∇ as a template, 108, 113

$(\mathbb{I}, \mathbb{E})$ a template, 112

$\mathbb{I}$ nodes of a template, 112

$\mathbb{E}$ edges of a template, 112

(Bij) a property on an adjunction, 164

(Nat) a property on an adjunction, 164

$- \dashv -$ adjunction
  between categories, 149
  between posets, 17

$f[\cdot]$ direct image across f, 80

$f^{\leftarrow}(\cdot)$ inverse image across f, 80

$\mathcal{LS}$ poset of lower sections of a poset S, 152, 194

$\widehat{S}$ presheaves on a poset, 27

$\exists$ an endo-functor
  on $\boldsymbol{Pos}$, 197
  on $\boldsymbol{Set}$, 79

I an endo-functor
  on $\boldsymbol{Pos}$, 197
  on $\boldsymbol{Set}$, 79

$\forall$ an endo-functor
  on $\boldsymbol{Pos}$, 197
  on $\boldsymbol{Set}$, 79

Particular categories

$\boldsymbol{AGrp}$ abelian groups, 28

$\boldsymbol{Ch(Mod\text{-}R)}$ chain complexes over R, 28

$\boldsymbol{CMon}$ commutative monoids, 141

$\boldsymbol{Eqv}$ equivalence relations, 137

$\boldsymbol{Fld}$ fields and morphisms, 47

$\boldsymbol{Grp}$ groups and morphisms, 8

$\boldsymbol{Idm}$ integral domains and morphisms, 47

$\boldsymbol{Inf}$ complete posets and $\bigwedge$-morphisms, 192

$\boldsymbol{Inv}$ involution algebras, 154, 204

$\boldsymbol{Mod\text{-}R}$ right R-modules and morphisms, 13

$\boldsymbol{MON}$ monoid set categories and functors, 88

$\boldsymbol{Mon}$ monoids and morphisms, 9

$\boldsymbol{Pfn}$ sets and partial functions, 11

$\boldsymbol{Pno}$ peano structures, 14

$\boldsymbol{Pos}$ posets and monotone maps, 10

$\boldsymbol{Pos}^{\dashv}$ posets and poset adjunctions, 17

$\boldsymbol{Pos}^{pp}$ posets and projection pairs, 24

$\boldsymbol{Pre}$ presets and monotone maps, 10

$\boldsymbol{Pth}(\nabla)$ path category of a graph, 113

RelA sets and relations as arrows, 16
R-Mod left *R*-modules and morphisms, 13
Rng rings and morphism, 8
R-Set left *R*-monoid sets and morphisms, 13
Set sets and functions, 7
SetD sets with a distinguished subset, 14, 134
Set$_\perp$ pointed sets and point preserving functions, 15
Set-R right *R*-monoid sets and morphisms, 13, 203
Sup complete posets and $\bigvee$-morphisms, 192
Top topological spaces and continuous maps, 8
Top$_2$ hausdorff spaces and continuous maps, 42
Vect$_K$ vector spaces over a field K and linear transformations, 8

Terminology

abstract nonsense, 72
action
    of a monoid on a set, 13
adjunction
    between category, 149, 151
    between poset, 17
arrow category, 19
arrow-class, 4

balanced category, 40
bimorphism, 40
biproduct, 55
butty category, 26

cell
    of a diagram, 35
chain complex, 28
choice function, 131
cocone, 49
codomain of an arrow, 2
coequalizer, 57
cofree solution, 176
commuting diagram, 5, 35
compact open topology, 159
cone, 49
confluent poset, 143
contravariant, 74
coproduct
    of two objects, 50

counit
    of an adjunction, 150
covariant, 74

diagonal functor, 78, 118
diagram, 2, 5, 35
    commuting, 5, 35
directed
    poset, 143
    pre-ordered set, 117
directed graph, 22, 113
domain of an arrow, 2

edge of a directed graph, 113
element
    global, 46
epic, 37
    split, 39
equalizer, 57

final object, 45
forgetful functor, 76
free solution, 176
functor, 73, 74
    contravariant, 74
    covariant, 74
    forgetful, 76
    hom, 77
furnished set, 5, 8
furnishings, 5, 8

global element, 46
graph
    as a directed graph, 22
    morphism, 22
    used as a template, 113

hom-functor, 77
hom-set, 4
horizontal composition, 106

idempotent algebra, 204
    2-step, 204
initial object, 45
inverse pair
    of isomorphisms, 40
involution algebra, 154, 204
    2-step, 204
isomorphism, 40

kernel
    of a function
        as an equivalence relation, 60

locally compact space, 160
loop space, 160
lower section
    of a poset, 193
    of a preset, 24

make equal, 56
map
    as another name for arrow, 2
mediating arrow, 50, 57, 66, 121
mediator, 50, 57, 66, 121
monic, 37
    split, 39
monoid, 8, 202
monoid set, 202
monotone map, 10
morphism
    as another name for arrow, 2

natural
    for variation of, 90
    isomorphism, 91
    transformation, 91
naturally equivalent, 91
node of a directed graph, 113

object
    final, 45
    initial, 45
    terminal, 45
    zero, 46
opposite
    of a category, 32

parallel pair of arrows, 5
partial function, 11
posets, 10
    $\bigvee$-complete, 191
    $\bigwedge$-complete, 191
    and monotone maps, 10
    and poset adjunctions, 17
    and projection pairs, 24
presets, 10
presheaf
    on a category, 96
    on a poset, 27
product
    of two categories, 19
    of two objects, 50
pullback, 65
pushout, 65

restriction of scalars, 155, 222
retraction
    a kind of arrow, 39

section
    a kind of arrow, 39
    a kind of subset of a preset
        lower, 24
        upper, 81
semigroup, 8
sets
    with a distinguished subset, 14
slice category, 20
solution
    to a posed problem, 64
source of an arrow, 2
specialization order, 81
split
    epic, 39
    monic, 39
structured set, 8
suspension space, 160

target of an arrow, 2
template, 19, 113
terminal object, 45
thread, 131
transposition assignment, 151

unit
    of an adjunction, 150
universal solution
    to a posed problem, 66, 121
upper section
    of a poset, 193
    of a preset, 81
upwards directed pre-ordered set, 117

vertical composition, 106

wedge, 49

Yoneda completion, 96

zero object, 46